# The Cooling of Neutron Stars - Neutrino Emissions from Neutron Stars with a Special Focus on the Direct Urca Process

Ronny Kjelsberg

2012

The image on the front cover is from PCModer.com and is licensed under a Creative Commons 3.0 license

All other contents Copyright ©2012 Lulu.com and Ronny Kjelsberg

All rights reserved

ISBN: 978-1-300-00864-4

*"...and it is feared that the French public, always impatient to come to a conclusion, eager to know the connections between general principles and the immediate questions that have aroused their passions, may be disheartened because they will be unable to move on at once.*

*That is a disadvantage I am powerless to overcome, unless it be by forewarning and forearming those readers who zealously seek the truth. There is no royal road to science, and only those who do not dread the fatiguing climb of its steep paths have a chance of gaining its luminous summits."*

*Karl Marx in the preface to the french edition of Capital (Progress Publishers, Moscow, 1954).*

# Forword

This book is mainly focused around neutrino emissions from neutron star cores, with an overweight on the direct Urca[1] process. It begins with a general introduction to neutron stars, the physics around compact objects and the historical development of neutron star physics. Furthermore i contains an overview of general relativity theory and the so-called TOV-equation derived by Tolman, Oppenheimer and Volkoff, to give an insight into the physics behind the development of an equation of state. Then we briefly look at the composition of the various equations of state, before we examine the consequences an inclusion of hyperons have on the equation of state, and the interactions which then must be included.

Moreover, given an overview of the various cooling processes, in which we will ourselves in the direct Urca process, which constitutes the main work of this task. We make calculations on the neutrino emissivity from the direct Urca process and makes plots of matter composition, chemical potential and nøytrinoemissivitet from the process with different interactions included. Finally, the results are discussed, and the direct Urca process is compared with other cooling processes.

This book is based on the work I did as a part of my cand.scient.-thesis at the Department of Physics at the Norwegian University of Science and Technology under the guidance of Prof.. Morten Hjorth-Jensen, University of Oslo. I would therefore especially like to thank Prof. Morten Hjorth-Jensen for good cooperation, and a good academic guidance during that work. Without his assistence the thesis, and this book, would not have seen the light of day. Furthermore, I wish also to thank my original supervisor, Prof.. Erlend Østgaard at NTNU which unfortunately was not able to complete this work with me, and I wish to thank my internal supervisor at NTNU, Professor

---

[1] The acronym URCA comes from the physicist George Gamow, and was the name of a casino in Rio de Janeiro. The casino was said to be effective in removing money from the pockets of its visitors. According to Gamows russian dialect, urca can also mean a pickpocket.

Sigmund Waldenstrøm for help with navigating within the bureaucracy at NTNU.

In addition, I must also give a general thanks to my fellow students for both the more and the less useful academic and extra-curricular discussions. In particular, I thank Sverre G. Johnsen and Anette Wrålsen for reading the original thesis and giving useful comments. Moreover, I thank various old and not quite as old men in the fifth floor of the the science building "Realfagbygget" at NTNU, who sometimes have been helpful in giving me little nudge in the right direction. I would also like to thank Rolf G. Lunder for useful programming tips. Finally, I will give a general thanks to all those who have given me the inspiration and knowledge during my education, and I have got to thank the website Google translate, without which this translation into english would have been much more time-consuming.

# Contents

Forword . . . . . . . . . . . . . . . . . . . . . . . . . . . . . . . . . . . . . . i

**1 Introduction to neutron star physics**     **1**

    1.1 Neutron stars . . . . . . . . . . . . . . . . . . . . . . . . . . . 1
    1.2 Neutron production . . . . . . . . . . . . . . . . . . . . . . . . 2
    1.3 The history of neutron star physics . . . . . . . . . . . . . . . 4
    1.4 Pulsars . . . . . . . . . . . . . . . . . . . . . . . . . . . . . . . 6
       1.4.1 The observational basis of neutron stars . . . . . . . . 6
       1.4.2 Identification of the relationship between neutron stars and supernovae . . . . . . . . . . . . . . . . . . . . . . . 7
       1.4.3 Why pulsars are neutron stars . . . . . . . . . . . . . . 8

**2 Equation of state and observables**     **9**

    2.1 The theory of general relativity and the TOV-equation . . . . 9
       2.1.1 General relativity . . . . . . . . . . . . . . . . . . . . . 9
       2.1.2 The TOV equation . . . . . . . . . . . . . . . . . . . . 19
    2.2 The equation of state and neutron star observables . . . . . . 21
    2.3 The Equation Of State . . . . . . . . . . . . . . . . . . . . . . 23

**3 The composition of equations of state**     **25**

    3.1 Ideal Fermi gas of neutrons . . . . . . . . . . . . . . . . . . . . 25
    3.2 Quark matter . . . . . . . . . . . . . . . . . . . . . . . . . . . 28
    3.3 Superfluid baryon matter . . . . . . . . . . . . . . . . . . . . . 29
    3.4 Kaon condensation . . . . . . . . . . . . . . . . . . . . . . . . 30
    3.5 Pion condensation . . . . . . . . . . . . . . . . . . . . . . . . . 31
    3.6 Phase Transitions . . . . . . . . . . . . . . . . . . . . . . . . . 32

# 4 Hyperon matter — 35

- 4.1 Description of compact matter with hyperons . . . . . . . . . . 35
- 4.2 Parameterization of the equation of state for nuclear matter . 39
- 4.3 Hyperon matter . . . . . . . . . . . . . . . . . . . . . . . . . . 42

# 5 Cooling processes — 51

- 5.1 Cooling processes in neutron stars . . . . . . . . . . . . . . . . 51
- 5.2 Direct Urca . . . . . . . . . . . . . . . . . . . . . . . . . . . . 53
  - 5.2.1 Direct Urca processes . . . . . . . . . . . . . . . . . . . 53
  - 5.2.2 Direct Urca with nucleons . . . . . . . . . . . . . . . . 54
  - 5.2.3 Direct Urca with muons . . . . . . . . . . . . . . . . . 56
  - 5.2.4 Direct Urca with hyperons . . . . . . . . . . . . . . . . 56

# 6 Neutrino emissivity — 59

- 6.1 Neutrino emissivity from the Direct Urca process . . . . . . . 59
- 6.2 A more exact calculation of the neutrino emissivity . . . . . . 64

# 7 Results — 73

# 8 Conclusion — 85

# A Program for calculating the emissivity — 87

# B Calculations in Maple — 101

# C Tables — 103

# Chapter 1

# Introduction to neutron star physics

## 1.1 Neutron stars

Compact objects are the end points of the development of a star. When the fusion of hydrogen into helium can no longer maintain the radiation pressure of the star, and the equilibrium that exists between this and the gravitational force is disturbed, the star is pulled together by gravity until reaching a temperature that allows the egnition of another nuclear process, which thus creates a new equilibrium. Eventually the star has however completely run out of nuclear fuel, and it is pulled together into a compact star; a white dwarf, a neutron star or a black hole. If the core mass is smaller than the so-called Chandrasekhar mass (about 1.4 $M_\odot$, where $M_\odot$ is the solar mass), degenerate electrons create a pressure that balances gravity, and we get a white dwarf. If the core mass is larger than the Chandrasekhar mass, the electrons become so relativistic that hydrostatic equilibrium is impossible, and the mass collapse continues until it is halted by nuclear forces. At this point the electrons have been captured through the process

$$e^- + p \rightarrow n + \nu_e \;, \tag{1.1}$$

so-called inverse $\beta$-decay, and the star ends up with the same density as the nucleus [**Ph99**]. This is a neutron star.

Under normal circumstances, the most stable form of nuclear matter is the one that is close to $^{56}$Fe. Less massive cores have a higher proportion of nucleons at the surface, and more massive ones get a large repulsion between protons. This changes when the temperature becomes so high that the electrons become relativistic. When the electrons get an energy greater than

1.3 MeV (the mass-energy difference between a proton and a neutron), we can get a $\beta$-decay in $^{56}$Fe, and we get the formation of more neutron rich cores,

$$e^- + {}^{56}\text{Fe} \rightarrow {}^{56}\text{Me} + \nu_e \ . \tag{1.2}$$

This process can continue with the production of $^{56}$Cr, and so on, as the density increases.

Electron capture using inverse $\beta$-decay occurs very rapidly when the core density exceeds $10^{14}$ kg m$^{-3}$. Neutrinos interact only to a limited degree with the matter in the star, and thus leave the star, bringning with them the energy that was originally stored by the degenerate electrons. When the pressure of electrons is lost, the star collapses rapidly and energy is transported away from the star in a violent eruptions of electron-neutrinos. The collapse happens quickly and takes place virtually without resistance until we reach a density comparable to the density of nuclear matter.

Analogous to the Chandrasekhar limit for white dwarfs, neutron stars can not have a mass greater than a certain critical limit. Since we have such strong gravitational fields, general relativity has to be taken into consideration in the calculation of this limit, and the maximum mass is difficult to calculate as there is uncertainty abot the compressability of neutron star matter at high densities. The limit is assumed to be about 3 M$_\odot$, and almost certainly less than 5 M$_\odot$. Above this limit, it is assumed that one can get the formation of black holes [**Ph99**].

## 1.2 Neutron production

As mentioned above, neutron stars are formed in a process where high pressure makes the electrons and protons in the star-matter join to form neutrons. We will now look more closely at this process.

If the electrons in a plasma have sufficient energy, they can join with protons and form neutrons. If $m_n$ and $m_p$ are the masses of respectively the neutron and proton, the electron must be as mentioned above have the total energy $E_{tot} = E^* > c^2 (m_n - m_p)$. At low densities the neutron desintegrere back to a proton-electron pair within 11 minutes, where the electron will have the total energy $E^*$ and a kinetic energy $E^*_{kin} = E^* - m_e c^2$, where $m_e$ is the electron mass [**KW90**]. If the gas is completely degenerate, and the phase space is filled up to the Fermi energy $E_F$, we can however end up with a different situation. If the Fermi energy $E_F$ exceeds $E^*_{kin}$, the electrons emitted in the disintegrations will not have enough energy to find a free space in the phase space, and the neutrons therefore can not desintegrate.

The Fermi sea thus has stabilized the neutrons. To find out more about under which conditions this happens, we can take alook at the equation for total energy,

$$E_{tot} = \frac{m_e c^2}{\sqrt{1 - v^2/c^2}} = m_e c^2 \sqrt{1 + \frac{p^2}{m_e^2 c^2}} \quad . \tag{1.3}$$

This equation we transform into

$$p = \frac{1}{c}\sqrt{E^2 + m_e^2 c^4} \quad . \tag{1.4}$$

If we set $E = E_{kin} + m_e c^2 = E_F + m_e c^2 = c^2 (m_n - m_p) = 1.294 \cdot 10^6$ eV we can find the corresponding Fermi momentum $p_F$ from Equation (1.4) above, and thus further discover that

$$x = \frac{p_F}{m_e c} \approx 2.2 \quad . \tag{1.5}$$

When we have a completely degenerate electron gas, the state where all electrons have the lowest energy without violating the Pauli principle will be the one where all the possible values in the phase space is occupied up to the specific momentum $p_F$. All places above $p_F$ will then be available. We get

$$\begin{aligned} f(p) &= \frac{8\pi p^2}{h^3} \quad \text{for } p \leq p_F \; , \\ f(p) &= 0 \quad \text{for } p \leq p_F \; , \end{aligned} \tag{1.6}$$

and the total number of electrons in the volume $dV$ will then be given by [**KW90**]

$$n_e dV = dV \int_0^{p_F} \frac{8\pi p^2 dp}{h^3} = \frac{8\pi}{h^3} p_F^3 dV \quad . \tag{1.7}$$

We then use $\rho = \mu_e m_u n_e$, where $m_u$ is an unit of mass $u$, and with $\mu_e = 2$ we get $\rho \approx 2{,}4 \cdot 10^7$ g cm$^{-3}$. If a gas of protons and electrons is compressed in this way to a density greater than this value, the gas will undergo a transition to a neutron gas, called neutronization.

For star matter, the situation is somewhat more complicated. At sufficiently high densities the matter contains heavier nuclei, and not just protons. Cores capture the electrons through the inverse $\beta$−decay, and become neutron rich isotopes. As the neutrons in the core are degenerate and the new neutrons must be lifted up to an energy level above the Fermi energy, this requires electron energies that are much higher than those estimated above. Similarly, we then need higher densities in the plasma to provide electrons with sufficient energy. When the cores then become too neutron rich they begin to break up, and they then release free neutrons. This so-called "neutron drip" starts as the density $\rho_{\text{drip}} \approx 4 \cdot 10^{11}$ g cm$^{-3}$.

Now let us briefly reflect on the impact of this on the equation of state. Up to $\rho_{\text{drip}}$, the total pressure of relativistic electrons will be $P \approx P_e$. When $\rho$ then continue to rise, the number density $n_e$ increases less than proportional to $\rho$ due to the electrons capture. The pressure therefore increases with less than $\rho^{4/3}$. At a continuously increasing $\rho$, neutrons will become increasingly degenerate, while the interaction between neutrons will become important. The details surrounding the equation of state are very uncertain, and depends, for example, on the properties of particles that are not as well known. We will get back to the equation of state in Chapter 2.

## 1.3 The history of neutron star physics

After the discovery of the neutron in 1932, Baade and Zwicky already in 1934 suggested [**BZ34**] that neutron stars could exist, stars with a very high density and small radius, who were very strongly bound by gravitational forces. These stars should consist of very closely packed neutrons. They also suggested that supernovae could be the transition from normal stars to these neutron stars.

The first calculations of neutron star models were conducted by Oppenheimer et al [**OV39**] in 1939. They believed then that neutron star matter consisted of an ideal gas of highly energetic, free neutrons. They used the equation of state of a relativistic degenerate Fermi gas and Einstein's field equations of general relativity theory to derive an equation for hydrostatic equilibrium. The work of this period was inspired by the idea that neutron cores in massive normal stars could be a source of the stars energy. This motivation, however, quickly disappeared when scientists later got an insight into the details of the nuclear fusion. Work with neutron stars was also given lower priority since the thermal radiation from them, because of their limited size, would be too small to be observed with optical telescopes across astronomical distances.

The discovery made by Giacconi et al [**GGP62**] in 1962, however, generated new interest in neutron stars. This discovery of cosmic X-ray sources that could not be ordinary stars, awakened new interest in the topic of neutron stars. It was proposed that these X-ray sources were young, hot neutron stars. The identification of quasars in 1963 created additional interest, since the large red shift of spectral lines that was observed, could be due to gravitational effects at the surface of a compact object. It turned out however that the largest red shift observed for quasars exceeded the maximum red shift from a stable neutron star. (The current scientific consensus says that a quasar is a compact region in the center of a massive galaxy, which surrounds a central super massive black hole.)

The theoretical work concentrated on the equations of state and neutron star models, equilibrium properties of compact stars and star collapse. In spite of all this research on neutron stars, most physicists and astronomers however, did not take the possibility of neutron stars (not to mention black holes) very seriously. The reason was probably the divergence neutron star physics represented from the known physics at the time. This changed however when a group of astronomers at Cambridge discovered pulsars in 1967.

In 1967, Hewish and others observed [**HBP68**] an astronomical object that sent out radio waves in periodic pulses. This had a great significance for the research into compact stars. The existence of stars that were more compact than the white dwarf was already theoretically described, and it was suggested that such objects could be created in a supernova explosion. It was also suggested that they initially would be rapidly rotating stars with an extremely strong magnetic field, and that a rotating neutron star could be the energy source of, for example, the Crab Nebula. A magnetic dipole model could convert the rotational energy of a neutron star to electromagnetic radiation and particle motions in the surrounding nebula.

The identification of pulsars as neutron stars nevertheless was not made until Gould in 1968 [**G68**] suggested that pulsars were rotating neutron stars with magnetic fields of an order of magnitude of $10^{12}$ Gauss. The existence of such objects could account for the remarkable stability that was observed for the pulse period, and Gould predicted a small increase of the period as the pulsar slowly lost rotational energy. This prediction was confirmed when such an increase was observed in the pulsar in the Crab Nebula, and this confirmation led to a general acceptance of the neutron star model. The discovery made in 1968 of the Crab and Vela pulsars - pulsars located in the remnants of supernovae - indicated that the assumption that neutron stars were created by supernova explosions was correct. The Crab Nebula, for instance, is the remnants of a supernova explosion that was observed by Chinese astronomers in 1054.

Theoretical work on neutron stars was further stimulated after the Uhuru satellite in 1971 discovered pulsating, compact X-ray sources. It is believed that these X-ray sources are neutron stars in binary systems which get a mass transfer - a so-called "accretion" - from an ordinary companion star. The first binary pulsar was discovered by Hulse and Taylor [**HT75**] in 1975, and such systems make it possible to measure the mass of a neutron star, and to examine the existence of radiation due to the large gravity.

## 1.4 Pulsars

### 1.4.1 The observational basis of neutron stars

It is therefore believed that pulsars are powerfully magnetized neutron stars, and theythus make up much of the observation basis for the theory behind neutron stars. An electromagnetic ray is emitted along the magnetic axis, with an angular distribution of typically about ten degrees. If the star has an angular misalignment between the rotational axis and the magnetic axis, the rotating object will look like a beacon and send out its radiation in the cones. If we on Earth are close to the axis of the star, within the spacial angle where the cone is sent out, we may, if it is not too distant, observe it as a pulsating radiation source. The frequency spectrum of the emitted signal from some pulsars, such as the Crab pulsar is huge, and can range from radio waves through optical light, X-rays and $\gamma$-radiation. Most pulsars, including all known millisecond pulsars, have however been observed only at radio wave frequencies.

Since Hewish et al detected the first in 1967, now over 1100 pulsars are detected. Some, such as the Crab and Vela pulsars, were discovered soon after Hewish' original discovery. Hundreds pulsarer were discovered in large searchs during the 70's, while a couple of hundred more have been discovered during the past years [G100].

The periods of the pulsars discovered to date, range from milliseconds to seconds, with an average period of approximately 0.7 seconds. The fastest one measured, PSR 1937+21, has got a period of 1.56 ms [G100].

Since the magnetic axis only randomly will coincide with the axis of rotation, the rotation itself causes the star to loose rotational energy due to the torque of the magnetic dipole. The rotational speed thus decreases over time, and, although this happens very slowly, one must not only give the period, but also the time of measurement, to give an accurate representation of the measurement of the period of a pulsar. This slow loss of angular momentum does not mean that the energy loss is small. The Crab Nebula pulsar loses energy at the rate of $2.5 \times 10^{44}$ MeV/s, but it has a rotational energy of $4 \times 10^{55}$ MeV.

It seems that there are two types of pulsars - two different populations. One type has a period that varies around approximately 0.7 s, and includes the majority of known pulsarer. These pulsars are usually isolated stars with strong magnetic fields of $\sim 10^{12}$ - $10^{13}$ Gauss. These are often referred to as canonical pulsars. The second population, millisecond pulsars, has, as the name suggests, a shorter period, which is in the range $10^{-3}$ - $10^{-2}$ s. More than halve of these are found in binary systems, and they all have weak magnetic fields of $\sim 10^8$ - $10^9$ Gauss. Although there has thus far been observed fewer millisecond pulsars, is believed that the galaxy contains a population of those who are of the same magnitude as the population of canonical pulsars.

It is believed that neutron stars are born with relatively short periods. The rotation will be spun up due to preservation of the angular momentum during the collapse to the core of the original star. The young Crab pulsar with a period of 1/30 s is perhaps a typical example of such a newborn neutron star. Such pulsars then evolve for a few million years, and get their periods increased their by the loss of rotational energy described above. During this period neutron stars are pulsars, but not necessarily visible to us here on Earth. This is of course dependent on the distance, magnitude of radiation, and whether the earth actually lies within the cone of radiation the pulsar sends out.

The radiation mechanism is poorly known, but at a critical point described by a product of the magnetic field strength and period, the cone beam stops. When the period increases, the radiation will cease at this point. The larger the field strength of the star, the longer a period is tolerated before this point is reached.

The star thus over time begins to rotate more slowly, from a loss of angular momentum due to the radiation. As the pulsar is slowing down its rotational pace, its radiation will also lose intensity. Pulsars therefore move slowly towards the critical point where radiation ceases.

## 1.4.2 Identification of the relationship between neutron stars and supernovae

We believe that neutron stars are created in supernova explosions of stars with mass of about 8 solar masses or more. Yet there are few cases where we certainly can identify pulsars with specific supernovae. Between 7 and 23 such identifications seem certain, depending on how stringent one should be on the terms used to make this identification [Gl00]. The most obvious problem with such identification is that the remains of a supernova explosion and a pulsar can be located on the same line of sight from Earth, but at very different distances. Convincing identifications have been found especially among young pulsars. This is easy to understand when one knows that pulsars move at high speeds, and that older pulsars thus will have moved a long way from its origin. One does not know the cause of this great speed, on average, about 450 km / s, but the compact neutron star will not be slowed down by the interstellar medium as the remnants of the supernova explosion will. The pulsar thus will move outside and away from the other remnants of the original star. Another factor that makes observations difficult, is the ratio of strength between supernova remnants and pulsars. As long as supernova remnants are visible, they are bigger and stronger radiation sources than the pulsars, and thus can be seen at long distances. This whole issue is fairly complicated, and those who work in this field have only recently begun to gain a more complete understanding of it.

### 1.4.3 Why pulsars are neutron stars

Ordinary stars have magnetic fields of about 100 Gauss, and they rotate. When the stars then collapse, both the rotation and the magnetic field strength are scaled up by the conservation laws of the magnetic flux and the angular momentum. After the massive explosion that initiates the formation of a pulsar, there is no reason why the magnetic axis and rotational axis should coincide. The magnetic field will typically get a value of approximately $10^{12}$ Gauss, and it exerts a strong torque on the rotating star. Together with a "wind" of relativistic particles, it is thought to be the reason of the gradual increase of the period. The change in period is typically an order of three seconds in $10^5$ years.

A rotating star with a mass 1 $M_\odot$, and diameter of $\sim$10 km, can contain a rotational energy of $\sim 10^{55}$ MeV. Such amounts of energy are necessary to explain the enormous amont of radiation mentioned in Chapter 1.4.1.

# Chapter 2

# Equation of state and observables

## 2.1 The theory of general relativity and the TOV-equation

The so-called TOV-equation was derived by Tolman, Oppenheimer and Volkoff, and has a central position in the calculation of mass and radius of neutron stars. The TOV-equation is a general-relativistic equation for hydrostatic equilibrium. The calculations thus assume that the neutron stars are in hydrostatic equilibrium, ie, that the gravitational forces are in equilibrium with the radiation pressure.

We will here first find the so-called Schwarzchild-metric using general relativity theory with gravitational forces. We derive the TOV-equation using this and Einstein's field equations. First we begin with an introduction to general relativity theory. For a better and more extensive introduction we can refer to for example [BG99] or [Be89].

### 2.1.1 General relativity

From special relativity we know that the laws of physics should look the same in all inertial systems. Yet we often have cases where we want to look at systems that are not inertial systems. A neutron star, for example, can rotate at a speed that is sufficiently high that matter at the equator is moving at speeds near the speed of light. At the same time have the surrounding area of a neutron star is influenced by strong gravitational forces, and it is not immediately clear how to treat such a system based on special relativity theory. It was Einstein who was able to formulate a theory for such situations - the theory of general relativity - which can also describe accelerated systems and systems with gravity. We will now look more closely at this.

If we consider two events in a system of coordinates $(x_1, y_1, z_1, t_1)$ and $(x_2, y_2, z_2, t_2)$, it is defined an invariant interval between these:

$$\Delta s^2 = c^2 \Delta t^2 - \Delta x^2 - \Delta y^2 - \Delta z^2 , \qquad (2.1)$$

which infinitesimally becomes:

$$ds^2 = c^2 dt^2 - dx^2 - dy^2 - dz^2 . \qquad (2.2)$$

The equation above, can be written on the form

$$ds^2 = \eta_{\mu\nu} dx^\mu dx^\nu , \qquad (2.3)$$

where

$$x^0 = ct, \quad x^1 = x, \quad x^2 = y, \quad x^3 = z , \qquad (2.4)$$

and

$$\eta_{\mu\nu} = \begin{pmatrix} 1 & 0 & 0 & 0 \\ 0 & -1 & 0 & 0 \\ 0 & 0 & -1 & 0 \\ 0 & 0 & 0 & -1 \end{pmatrix} , \qquad (2.5)$$

where $\eta_{\mu\nu}$ is called the Minkowski-metric. There is no universal definition of this, so you can often be presented with versions of the metric in which all components (2.5) have changed their sign. No physical results will naturally depend on such a convention, but some intermediate results will differ.

The Greek summation indices $\mu$ and $\nu$ in the metric has values from 0 to 3, so that for example, $\eta_{00} = 1$ and $\eta_{11} = -1$. A summation index stands once with a high index and once with a low index in an expression, then meaning summation from 0 to 3 The four-vectors $x^\mu$ (high index) and $x_\mu$ (low index) are called respectively contravariant and covariant, and they are defined as follows:

$$x_\mu = \eta_{\mu\nu} x^\nu . \qquad (2.6)$$

Meaning

$$\begin{aligned} x_0 &= x^0 , \\ x_1 &= -x^1 , \\ x_2 &= -x^2 , \\ x_3 &= -x^3 . \end{aligned} \qquad (2.7)$$

We can then write the interval on the simple form

$$ds^2 = dx_\mu dx^\mu \ . \tag{2.8}$$

We have now considered the metric of a flat space without a gravity field. With a gravitational field, we find the following expression for the line element:

$$ds^2 = g_{\mu\nu} dx_\mu dx^\mu \ . \tag{2.9}$$

We assume symmetry:

$$g_{\nu\mu} = g_{\mu\nu} \ , \tag{2.10}$$

and we assume that $g_{\mu\nu}$ is approaching $\eta_{\mu\nu}$ when the gravitational field $G$ approaches 0,

$$\lim_{G \to 0} g_{\mu\nu} = \eta_{\mu\nu} \ . \tag{2.11}$$

We than assume a static (time-independent) field, in which all $g_{\mu\nu}$ are time-independent and all $g_{0i} = 0$. The line element from (2.9) can then be reduced into:

$$ds^2 = g_{00} c^2 dt^2 + g_{ij} dx^i dx^j \ , \ (\text{der } i,j \in \{1,2,3\}) \ . \tag{2.12}$$

Around a static spherically symmetric mass, we naturally have a spherically symmetric gravitational field, and it is therefore convenient to introduce the spherical coordinates: $x^1 = r$, $x^2 = \theta$, $x^3 = \varphi$. The relationship between Cartesian and spherical coordinates are:

$$\begin{aligned} x &= r \sin\theta \cos\varphi \ , \\ y &= r \sin\theta \sin\varphi \ , \\ z &= r \cos\theta \ . \end{aligned} \tag{2.13}$$

We also want to have a system that is symmetric with respect to rotation. In this case, the line element must be independent of the transformation $d\theta \leftrightarrow -d\theta$, or $d\varphi \leftrightarrow -d\varphi$. Thus we can have no terms of the type $d\theta d\varphi$, $d\varphi d\theta$ etc. We can then write the line element on the following form [**Rø01**, **Mø83**]:

$$ds^2 = A(r) c^2 dt^2 - B(r) dr^2 - r^2 d\theta^2 - r^2 \sin^2\theta d\varphi^2 \ . \tag{2.14}$$

$A(r)$ and $B(r)$ are arbitrary functions. These can be written as exponential functions [**ABS75**], and we will then get the expression

$$ds^2 = e^{\nu(r)} c^2 dt^2 - e^{\lambda(r)} dr^2 - r^2 d\theta^2 - r^2 \sin^2\theta d\varphi^2 \ , \tag{2.15}$$

where the parameters $\nu(r)$ and $\lambda(r)$ are arbitrary functions to be determined later. The metric tensor $g_{\mu\nu}$, will then become:

$$g_{\mu\nu} = \begin{pmatrix} e^{\nu(r)} & 0 & 0 & 0 \\ 0 & -e^{\lambda(r)} & 0 & 0 \\ 0 & 0 & -r^2 & 0 \\ 0 & 0 & 0 & -r^2\sin^2\theta \end{pmatrix} . \quad (2.16)$$

To find the functions $\nu(r)$ and $\lambda(r)$, we use the Einstein field equations. Relativity theory is mathematically quite extensive, and to simplify, we will use the Christoffel symbols, who are defined by [Mø83, Dy91]:

$$\Gamma^{\mu}_{\alpha\beta} \stackrel{def}{=} \frac{1}{2}g^{\mu\nu}\left(\frac{\partial g_{\nu\alpha}}{\partial x^{\beta}} + \frac{\partial g_{\beta\nu}}{\partial x^{\alpha}} - \frac{\partial g_{\alpha\beta}}{\partial x^{\nu}}\right) , \quad (2.17)$$

and have the property $\Gamma^{\mu}_{\alpha\beta} = \Gamma^{\mu}_{\beta\alpha}$. When we then insert for the metric tensor from (2.16), we get the expressions:

$$\begin{aligned}\Gamma^{0}_{00} &= \frac{1}{2}g^{00}\left(\frac{\partial g_{00}}{\partial x^{0}} + \frac{\partial g_{00}}{\partial x^{0}} - \frac{\partial g_{00}}{\partial x^{0}}\right) = \frac{1}{2}g^{00}\frac{\partial g_{00}}{\partial x^{0}} = \frac{1}{2}e^{-\nu(r)}e^{\nu(r)}\frac{\partial \nu}{\partial (ct)} \\ &= \frac{1}{2c}\frac{\partial \nu}{\partial t} ,\end{aligned} \quad (2.18)$$

$$\Gamma^{1}_{00} = \frac{1}{2}g^{11}\left(\frac{\partial g_{10}}{\partial x^{0}} + \frac{\partial g_{01}}{\partial x^{0}} - \frac{\partial g_{00}}{\partial x^{1}}\right) = \frac{1}{2}g^{11}\frac{\partial g_{00}}{\partial x^{1}} = \frac{1}{2}e^{-\lambda(r)}e^{\nu(r)}\frac{\partial \nu}{\partial r} , \quad (2.19)$$

and correspondingly

$$\begin{aligned}\Gamma^{1}_{11} &= \frac{1}{2}\frac{\partial \lambda}{\partial r} , \\ \Gamma^{1}_{22} &= -re^{-\lambda(r)} , \\ \Gamma^{1}_{33} &= -r\sin^2\theta\, e^{-\lambda(r)} , \\ \Gamma^{2}_{33} &= -\sin\theta\cos\theta , \\ \Gamma^{0}_{01} &= \Gamma^{0}_{10} = \frac{1}{2}\frac{\partial \nu}{\partial r} , \\ \Gamma^{2}_{12} &= \Gamma^{2}_{21} = \frac{1}{r} , \\ \Gamma^{3}_{13} &= \Gamma^{3}_{31} = \frac{1}{r} , \\ \Gamma^{3}_{23} &= \Gamma^{3}_{32} = \cot\theta .\end{aligned} \quad (2.20)$$

The Riemann tensor [**Rø01, BG99**] gives us a useful relationship with the Christoffel symbols:

$$R^{\alpha}_{\beta\mu\nu} \stackrel{def}{=} \frac{\partial \Gamma^{\alpha}_{\beta\nu}}{\partial x^{\mu}} - \frac{\partial \Gamma^{\alpha}_{\beta\mu}}{\partial x^{\nu}} + \Gamma^{\alpha}_{\sigma\mu}\Gamma^{\sigma}_{\beta\nu} - \Gamma^{\alpha}_{\sigma\nu}\Gamma^{\sigma}_{\beta\nu} \ . \tag{2.21}$$

The Riemann tensor appears to be complicated, but thanks to a large number of symmetries the number of independent components is reduced from what initially appears to be $4^4 = 256$ to 20. These symmetries, we can summarize easyest for the tensor $R_{\alpha\beta\gamma\delta} = g_{\alpha\rho}R^{\rho}_{\beta\gamma\delta}$, which is formed by lowering the first index. We have a symmetry in the exchange of the first and second index pair:

$$R_{\alpha\beta\gamma\delta} = R_{\gamma\delta\alpha\beta} \ . \tag{2.22}$$

We have anti-symmetry in the exchanges:

$$R_{\alpha\beta\gamma\delta} = -R_{\beta\alpha\gamma\delta} = -R_{\alpha\beta\delta\gamma} \ , \tag{2.23}$$

and syclical properties:

$$R_{\alpha\beta\gamma\delta} + R_{\alpha\delta\beta\gamma} + R_{\alpha\gamma\delta\beta} = 0 \ . \tag{2.24}$$

By contraction of the first and the third index in the Riemann tensor, we find the Ricci tensor

$$R_{\beta\mu} = R^{\alpha}_{\beta\alpha\mu} \ . \tag{2.25}$$

The Ricci tensor has 16 components, of which 10 are independent. The Ricci tensor is symmetric over the exchange of the indices, and by contraction of the two indices we get a scalar, called the Ricci scalar, or the scalar curvature of the space,

$$R = g^{\beta\mu}R_{\beta\mu} = g^{\beta\alpha}g^{\mu\sigma}R_{\beta\mu\alpha\sigma} \ . \tag{2.26}$$

From the Ricci tensor and Ricci scalar, we can form another symmetric tensor with decreasing kovariant divergence. This is the Einstein tensor [**GB99**],

$$G_{\beta\mu} = R_{\beta\mu} - \frac{1}{2}g_{\beta\mu}R \ . \tag{2.27}$$

Through this we get a correlation between space curvature and the energy distribution in space [**Mø83, BG99**]:

$$R_{\beta\mu} - \frac{1}{2}g_{\beta\mu}R = \frac{8\pi G}{c^4}T_{\beta\mu} \ , \tag{2.28}$$

which are Einstein's field equations. Here $G$ is the universal gravitational constant, $c$ is the speed of light and $T_{\beta\mu}$ is the four-dimensional energy-momentum tensor. $T_{\beta\mu}$ is given by:

$$T_{\beta\mu} = \rho u_{\beta}u_{\mu} + P\left(\frac{u_{\beta}u_{\mu}}{c^2} - g_{\beta\mu}\right) \ , \tag{2.29}$$

where $\rho$ is the density, $P$ is pressure, and $u_\beta$ and $u_\mu$ are four-velocities. Now we can multiply (2.28) by $g^{\beta\mu}$, and use the condition

$$g^{\beta\mu} g_{\beta\mu} = \partial^\beta_\mu \, , \qquad (2.30)$$

where

$$\partial^\beta_\mu = \begin{cases} 1 & \text{for} \quad \beta = \mu \\ 0 & \text{ellers.} \end{cases} \qquad (2.31)$$

This gives us

$$g^{\beta\mu} R_{\beta\mu} - \frac{1}{2} g^{\beta\mu} g_{\beta\mu} R = \frac{8\pi G}{c^4} g^{\beta\mu} T_{\beta\mu}$$

$$R - \frac{1}{2}\left(g^{00} g_{00} + g^{11} g_{11} + g^{22} g_{22} + g^{33} g_{33}\right) = \frac{8\pi G}{c^4} T^\beta_\beta$$

$$R - \frac{1}{2}(4) R = \frac{8\pi G}{c^4} T^\beta_\beta$$

$$R = -\frac{8\pi G}{c^4} T^\beta_\beta \, , \qquad (2.32)$$

where $T = T^\beta_\beta$ is called the trail of the energy-momentum tensor. When we set (2.32) in to (2.28), we get

$$R_{\beta\mu} = \frac{8\pi G}{c^4}\left(T_{\beta\mu} - R_{\beta\mu} - \frac{1}{2} g_{\beta\mu} T\right) \, . \qquad (2.33)$$

We can now determin $g_{\beta\mu}$ by finding the expressions for $R_{\beta\mu}$, $T_{\beta\mu}$ and $T$, and inserting them into the equation.

First we try to determine the components of the Ricci tensor, $R_{\beta\mu}$. From Equation (2.25) we have that $R_{\beta\mu} = R^\alpha_{\beta\alpha\mu}$. This gives us:

$$\begin{aligned}
R_{00} &= R^0_{000} + R^1_{010} + R^2_{020} + R^3_{030} \, , \\
R_{11} &= R^0_{101} + R^1_{111} + R^2_{121} + R^3_{131} \, , \\
R_{22} &= R^0_{202} + R^1_{212} + R^2_{222} + R^3_{232} \, , \\
R_{33} &= R^0_{303} + R^1_{313} + R^2_{323} + R^3_{333} \, .
\end{aligned} \qquad (2.34)$$

We have expressed the Riemann tensor, by the Cristoffel symbols in Equation

(2.21). The Christoffel symbols that contribute, we find from (2.18)-(2.20):

$$R^0_{000} = \frac{d\Gamma^0_{00}}{dx^0} - \frac{d\Gamma^0_{00}}{dx^0} + \Gamma^0_{n0}\Gamma^n_{00} - \Gamma^0_{n0}\Gamma^n_{00} = 0 \; , \tag{2.35}$$

$$\begin{aligned} R^1_{010} &= \frac{d\Gamma^1_{00}}{dx^1} - \frac{d\Gamma^1_{01}}{dx^0} + \Gamma^1_{n1}\Gamma^n_{00} - \Gamma^1_{n0}\Gamma^n_{01} \\ &= \frac{d}{dr}\left(\frac{1}{2}\frac{d\nu}{dr}e^{(\nu(r)-\lambda(r))}\right) + \Gamma^1_{01}\Gamma^0_{00} - \Gamma^1_{00}\Gamma^0_{01} + \Gamma^1_{11}\Gamma^1_{00} - \Gamma^1_{10}\Gamma^1_{01} \\ &\quad + \Gamma^1_{21}\Gamma^2_{00} - \Gamma^1_{20}\Gamma^2_{01} + \Gamma^1_{31}\Gamma^3_{00} - \Gamma^1_{30}\Gamma^3_{01} \\ &= \frac{d}{dr}\left(\frac{1}{2}\frac{d\nu}{dr}e^{(\nu(r)-\lambda(r))}\right) - \Gamma^1_{00}\Gamma^0_{01} + \Gamma^1_{11}\Gamma^1_{00} \\ &= \frac{1}{2}\left(\frac{d^2\nu}{dr^2}e^{(\nu(r)-\lambda(r))} + \left(\frac{d\nu}{dr} - \frac{d\lambda}{dr}\right)\frac{d\nu}{dr}e^{(\nu(r)-\lambda(r))}\right) \\ &\quad - \frac{1}{2}\frac{d\nu}{dr}\frac{1}{2}\frac{d\nu}{dr}e^{(\nu(r)-\lambda(r))} + \frac{1}{2}\frac{d\nu}{dr}e^{(\nu(r)-\lambda(r))} \cdot \frac{1}{2}\frac{d\lambda}{dr} \\ &= e^{(\nu(r)-\lambda(r))} \cdot \left(\frac{1}{2}\frac{d^2\nu}{dr^2} - \frac{1}{4}\frac{d\nu}{dr}\frac{d\lambda}{dr} + \frac{1}{4}\left(\frac{d\nu}{dr}\right)^2\right) \; , \end{aligned} \tag{2.36}$$

$$\begin{aligned} R^2_{020} &= \frac{d\Gamma^2_{00}}{dx^0} - \frac{d\Gamma^2_{02}}{dx^0} + \Gamma^2_{n2}\Gamma^n_{00} - \Gamma^2_{n0}\Gamma^n_{02} \\ &= \frac{1}{2}\cdot\frac{1}{r}\frac{d\nu}{dr}e^{(\nu(r)-\lambda(r))} \; , \end{aligned} \tag{2.37}$$

$$\begin{aligned} R^3_{030} &= \frac{d\Gamma^3_{00}}{dx^3} - \frac{d\Gamma^3_{03}}{dx^0} + \Gamma^3_{n3}\Gamma^n_{00} - \Gamma^3_{n0}\Gamma^n_{03} \\ &= \frac{1}{2}\cdot\frac{1}{r}\frac{d\nu}{dr}e^{(\nu(r)-\lambda(r))} \; . \end{aligned} \tag{2.38}$$

If we put these results into the equations (2.34), we find $R_{00}$, $R_{11}$, $R_{22}$ and $R_{33}$:

$$R_{00} = e^{(\nu(r)-\lambda(r))}\left(\frac{1}{2}\frac{d^2\nu}{dr^2} - \frac{1}{4}\frac{d\nu}{dr}\frac{d\lambda}{dr} + \frac{1}{4}\left(\frac{d\nu}{dr}\right)^2 + \frac{1}{r}\frac{d\nu}{dr}\right) \; , \tag{2.39}$$

$$R_{11} = -\frac{1}{2}\frac{d^2\nu}{dr^2} + \frac{1}{4}\frac{d\nu}{dr}\frac{d\lambda}{dr} - \frac{1}{4}\left(\frac{d\nu}{dr}\right)^2 + \frac{1}{r}\frac{d\lambda}{dr} \; , \tag{2.40}$$

$$R_{22} = e^{-\lambda(r)}\left(-1 - \frac{1}{2}r\frac{d\nu}{dr} + \frac{1}{2}\left(r\frac{d\lambda}{dr}\right) + 1\right) \; , \tag{2.41}$$

$$R_{33} = R_{22}\sin^2\theta \; . \tag{2.42}$$

Now we will attempt to identify the components of the energy-momentum tensor, $T_{\beta\mu}$. To find $T_{\beta\mu}$, we must solve Equation (2.29):

$$T_{\beta\mu} = \rho u_\beta u_\mu + P\left(\frac{u_\beta u_\mu}{c^2} - g_{\beta\mu}\right) \quad . \tag{2.43}$$

The four-velocity is defined by

$$u^\nu u_\mu = g_{\mu\alpha} u^\nu u^\alpha = c^2 \quad , \tag{2.44}$$

and for a liquid at rest, we get the velocity component

$$u_\beta = (u_0, 0, 0, 0) \quad , \tag{2.45}$$

i.e. (2.44) gives us

$$g_{00} \cdot \left(u^0\right)^2 = c^2 \quad , \tag{2.46}$$

$$u_0 = g_{0\mu} u^\mu = g_{00} u^0 = \sqrt{g_{00}} \cdot c \quad . \tag{2.47}$$

We can then write the energy-momentum tensor as

$$\begin{aligned} T_{\beta\mu} &= \rho\left(u_0\right)^2 + P\left(\frac{(u_0)^2}{c^2} - g_{00} - g_{11} - g_{22} - g_{33}\right) \\ &= \rho c^2 g_{00} + P\left(g_{00} - g_{00} - g_{11} - g_{22} - g_{33}\right) \\ &= \rho c^2 g_{00} - P\left(g_{11} + g_{22} + g_{33}\right) \quad . \end{aligned} \tag{2.48}$$

This gives us the components

$$\begin{aligned} T_{00} &= \rho c^2 e^{\nu(r)} \quad , \\ T_{11} &= P e^{\lambda(r)} \quad , \\ T_{22} &= P r^2 \quad , \\ T_{33} &= P r^2 \sin^2\theta \quad . \end{aligned} \tag{2.49}$$

Finally, we now find the trace of the energy-momentum tensor, $T$. To determine this, we multiply Equation (2.29) by $g^{\beta\mu}$:

$$\begin{aligned} g^{\beta\mu} T_{\beta\mu} = T &= \rho g^{\beta\mu} u_\beta u_\mu + P\left(\frac{g^{\beta\mu} u_\beta u_\mu}{c^2} - 4\right) \\ &= \rho c^2 - 3P \quad . \end{aligned} \tag{2.50}$$

When we now have this information, we can solve Equation (2.33) with respect to $g^{\beta\mu}$, and thereby with respect to $\nu(r)$ and $\lambda(r)$,

$$R_{\beta\mu} = \frac{8\pi G}{c^4}\left(T_{\beta\mu} - \frac{1}{2} g_{\beta\mu} T\right) \quad . \tag{2.51}$$

Now we will insert the components for $R_{\beta\mu}$ from the equations (2.34), $T_{\beta\mu}$ from (2.49) and $T$ from (2.50). This gives us four equations [**Rø01**]:

$$e^{-\lambda}\left(\frac{1}{2}\frac{d^2\nu}{dr^2} - \frac{1}{4}\frac{d\nu}{dr}\frac{d\lambda}{dr} + \frac{1}{4}\left(\frac{d\nu}{dr}\right)^2 + \frac{1}{r}\frac{d\nu}{dr}\right) = \frac{8\pi G}{c^4}\left(\frac{\rho c^2}{2} + \frac{3P}{2}\right), \tag{2.52}$$

$$e^{-\lambda}\left(-\frac{1}{2}\frac{d^2\nu}{dr^2} + \frac{1}{4}\frac{d\nu}{dr}\frac{d\lambda}{dr} - \frac{1}{4}\left(\frac{d\nu}{dr}\right)^2 + \frac{1}{r}\frac{d\nu}{dr}\right) = \frac{8\pi G}{c^4}\left(\frac{\rho c^2}{2} - \frac{P}{2}\right), \tag{2.53}$$

$$e^{-\lambda}\left(-\frac{1}{r^2} - \frac{1}{2r}\left(\frac{d\nu}{dr} - \frac{d\lambda}{dr}\right)\right) + \frac{1}{r^2} = \frac{8\pi G}{c^4}\left(\frac{\rho c^2}{2} - \frac{P}{2}\right), \tag{2.54}$$

$$e^{-\lambda}\left(-\frac{1}{r^2} - \frac{1}{2r}\left(\frac{d\nu}{dr} - \frac{d\lambda}{dr}\right)\right) + \frac{1}{r^2} = \frac{8\pi G}{c^4}\left(\frac{\rho c^2}{2} - \frac{P}{2}\right). \tag{2.55}$$

Here we see that the last two equations are identical. By adding The two first equations, we get

$$e^{-\lambda}\left(\frac{1}{r}\left(\frac{d\nu}{dr} - \frac{d\lambda}{dr}\right)\right) = \frac{8\pi G}{c^4}\left(\rho c^2 + P\right). \tag{2.56}$$

From this we see that $\left(\frac{d\nu}{dr} - \frac{d\lambda}{dr}\right)$ will allways be positive or equal to zero. It is equal to zero only for $\left(\rho c^2 + P\right) = 0$, i.e. when $\rho = P = 0$. In this case the energy-momentum tensor $T_{\beta\mu}$ from Equation (2.29), will also be zero. When we thus add (2.55) and (2.56), we will find a solution for $\rho$:

$$\frac{8\pi G}{c^2}\rho = e^{-\lambda}\left(-\frac{1}{r^2} + \frac{1}{r}\frac{d\lambda}{dr}\right) + \frac{1}{r^2}. \tag{2.57}$$

If we on the other hand subtract (2.55) from (2.56), we will find a solution for $P$:

$$\frac{8\pi G}{c^2}P = -\frac{1}{r^2} + e^{-\lambda} + \left(\frac{1}{r^2} + \frac{1}{r}\frac{d\nu}{dr}\right). \tag{2.58}$$

We then subtract (2.53) from (2.55), and get an equation without neither $\rho$ or $P$:

$$\frac{1}{r^2}e^{-\lambda} = \frac{1}{r^2} - \frac{1}{4}\left(\frac{d\nu}{dr}\right)^2 + \frac{1}{4}\left(\frac{d\nu}{dr}\frac{d\lambda}{dr}\right) + \frac{1}{2r}\left(\frac{d\nu}{dr} + \frac{d\lambda}{dr}\right) - \frac{1}{2}\frac{d^2\nu}{dr^2}. \tag{2.59}$$

We are now left with three differential equations (2.57), (2.58) and (2.59) for the four functions $\nu(r)$, $\lambda(r)$, $P(r)$ and $\rho(r)$. If we multiply Equation (2.57) by $r^2$ we get:

$$e^{-\lambda}\left(r\frac{d\lambda}{dr} - 1\right) + 1 = \frac{8\pi G}{c^2}\rho r^2 ,$$

$$r\frac{d\lambda}{dr}e^{-\lambda} - e^{-\lambda} = \frac{8\pi G}{c^2}\rho r^2 ,$$

$$\frac{d}{dr}\left(re^{-\lambda}\right) = \frac{8\pi G}{c^2}\rho r^2 ,$$

$$e^{-\lambda} = 1 - \frac{8\pi G}{rc^2}\int_0^r \rho r^2 dr . \quad (2.60)$$

We then apply the expression of total mass, $M(r) = 4\pi \int_0^r \rho r^2 dr$, and end up with

$$e^{-\lambda} = 1 - \frac{2GM}{rc^2} . \quad (2.61)$$

We will now assume a spherically symmetrical gravitational field in vacuum. In this special case, the energy-momentum tensor $T_{\beta\mu}$ will give us 0. We will then get from (2.57) and (2.58)

$$e^{-\lambda}\left(\frac{1}{r^2} - \frac{1}{r}\frac{d\lambda}{dr}\right) - \frac{1}{r^2} = 0 , \quad (2.62)$$

$$e^{-\lambda}\left(\frac{1}{r^2} + \frac{1}{r}\frac{d\nu}{dr}\right) - \frac{1}{r^2} = 0 . \quad (2.63)$$

When we take the difference between these equations, and get

$$\frac{d\lambda}{dr} + \frac{d\nu}{dr} = 0 . \quad (2.64)$$

When we then integrate this equation, we get

$$\nu(r) = -\lambda(r) + C , \quad (2.65)$$

where C is an arbitrary constant. Moreover, we can set the condition that when $r \to \infty$, the metric tensor $g_{\mu\nu}$ must be reduced to $\eta_{\mu\nu}$, which means that $e^\lambda$ and $e^\nu \to 1$. From this we can conclude that $C = 0$, and thereby that

$$\nu(r) = -\lambda(r) . \quad (2.66)$$

Together with the expression for $e^{-\lambda(r)}$ from (2.61), we can insert this into the expression for the metric tensor $g_{\mu\nu}$ from (2.16). This gives us the Schwarzschild metric, the metric around a spherically symmetrical mass in a gravitational field.

$$g_{\mu\nu} = \begin{pmatrix} 1 - \frac{2GM}{rc^2} & 0 & 0 & 0 \\ 0 & -\left(1 - \frac{2GM}{rc^2}\right)^{-1} & 0 & 0 \\ 0 & 0 & -r^2 & 0 \\ 0 & 0 & 0 & -r^2 \sin^2\theta \end{pmatrix} . \quad (2.67)$$

The interval $ds^2$ can be written as

$$ds^2 = \left(1 - \frac{2GM}{rc^2}\right) c^2 dt^2 - \left(1 - \frac{2GM}{rc^2}\right)^{-1} dr^2 - r^2 \left(d\theta^2 + \sin^2\theta d\phi^2\right) . \quad (2.68)$$

## 2.1.2 The TOV equation

We will now use these results to derive the TOV equation. If we adopt the expression for $e^{-\lambda(r)}$ from (2.61), we can solve Equation (2.58) with respect to $\frac{d\nu}{dr}$, i.e.

$$\frac{8\pi G}{c^4} P = e^{-\lambda(r)} \left(\frac{1}{r^2} + \frac{1}{r}\frac{d\nu}{dr}\right) - \frac{1}{r^2} , \quad (2.69)$$

which gives us

$$\begin{aligned} \frac{d\nu}{dr} &= re^{\lambda(r)} \left(\frac{8\pi G}{c^4} P - \frac{1}{r^2} e^{-\lambda(r)} + \frac{1}{r^2}\right) \\ &= \left(\frac{8\pi G P r}{c^4} - \frac{1}{r}\left[1 - \frac{2GM}{c^2 r}\right] + \frac{1}{r}\right) / \left[1 - \frac{2GM}{c^2 r}\right] \\ &= \frac{2G\left(M + 4\pi r^3 P/c^2\right)}{c^2 r^2 \left(1 - 2GM/c^2 r\right)} . \end{aligned} \quad (2.70)$$

Now we will try to find the relationship between $\frac{d\nu}{dr}$ and $\frac{dP}{dr}$. We then derive (2.58):

$$-\frac{8\pi G}{c^4}\frac{dP}{dr} =$$
$$-\frac{2}{r^3} + \frac{d\lambda}{dr} e^{-\lambda(r)} \left(\frac{1}{r^2} + \frac{1}{r}\frac{d\nu}{dr}\right) - e^{-\lambda(r)} \left(-\frac{2}{r^3} + \frac{1}{r}\frac{d^2\nu}{dr^2} - \frac{1}{r^2}\frac{d\nu}{dr}\right)$$
$$= -\frac{2}{r^3} + \frac{1}{r} e^{-\lambda(r)} \left(\frac{1}{r}\frac{d\lambda}{dr} + \frac{d\lambda}{dr}\frac{d\nu}{dr} + \frac{2}{r^2} - \frac{d^2\nu}{dr^2} + \frac{1}{r}\frac{d\nu}{dr}\right) . \quad (2.71)$$

19

The we solve (2.59) with respect to $\frac{d^2\nu}{dr^2}$, and insert this expression into (2.71). We then get the expression

$$
\begin{aligned}
-\frac{8\pi G}{c^4}\frac{dP}{dr} &= -\frac{2}{r^3} + \frac{1}{r}e^{-\lambda(r)}\Big(\frac{1}{r}\frac{d\lambda}{dr} + \frac{d\lambda}{dr}\frac{d\nu}{dr} + \frac{2}{r^2} + \frac{1}{r}\frac{d\nu}{dr} \\
&\quad + \frac{2e^{\lambda(r)}}{r^2} - \frac{2}{r^2} + \frac{1}{2}\frac{d\nu}{dr}\frac{d\lambda}{dr} - \frac{1}{r}\frac{d\nu}{dr} - \frac{1}{r}\frac{d\lambda}{dr}\Big) \\
&= -\frac{2}{r^3} + \frac{1}{r}e^{-\lambda(r)}\left(\frac{1}{2}\left(\frac{d\nu}{dr}\right)^2 + \frac{1}{2}\frac{d\nu}{dr}\frac{d\lambda}{dr}\right) + \frac{2}{r^3} \\
&= e^{-\lambda(r)}\frac{1}{r}\left(\frac{d\nu}{dr} + \frac{d\lambda}{dr}\right)\frac{1}{2}\frac{d\nu}{dr} \\
&= \frac{8\pi G}{c^4}\left(\rho c^2 + P\right)\frac{1}{2}\frac{d\nu}{dr} \quad,
\end{aligned}
\tag{2.72}
$$

which finally gives us

$$
\frac{dP}{dr} = -\left(\rho c^2 + P\right)\frac{1}{2}\frac{d\nu}{dr} \quad. \tag{2.73}
$$

When we take and insert the expression for $\frac{d\nu}{dr}$ into this equation, we get

$$
\frac{dP}{dr} = -\frac{1}{2}\left(\rho c^2 + P\right)\frac{2G\left(M + 4\pi r^3 P/c^2\right)}{c^2 r^2\left(1 - 2GM/c^2 r\right)} \quad. \tag{2.74}
$$

This gives us

$$
\frac{dP}{dr} = -\frac{G\left(\rho c^2 + P\right)\left(M + 4\pi r^3 P/c^2\right)}{c^2 r^2\left(1 - 2GM/c^2 r\right)} \quad, \tag{2.75}
$$

which is the TOV-equation we were looking for. Together with the equation for mass-energy conservation

$$
\frac{dM}{dr} = 4\pi r^2 \cdot \rho(r) \tag{2.76}
$$

and the equation of state

$$
P = P(\rho) \quad, \tag{2.77}
$$

we then have three equations with three unknown variables $P(r)$, $\rho(r)$ og $M(r)$. Here $P(r)$ is the pressure, $\rho(r)$ the mass-energy density, $M(r)$ is the mass within $r$, $G$ is the universal gravitational constant and $c$ is the speed of light. We can then determine the total mass $M$ of a neutron star and the radius of the star would then be defined by

$$
P(R) = 0 \quad. \tag{2.78}
$$

## 2.2 The equation of state and neutron star observables

One of the most important aspects in the study of neutron stars is the equation of state, i.e. the functional dependence between pressure $p$ and density $\rho$, which describes the internal structure of the star.

The physics of compact objects such as neutron stars, gives us a unique opportunity to look at the interaction between nuclear processes and astrophysical observables, and by looking at neutron stars, we can gain insight into an environment remote from what we can observe on earth, see for example [ST83, Gl00]. To determine an equation of state for compact matter is central for the calculation of properties of neutron stars, and it determines both the mass limit, and the relationship between mass and radius for these stars. Equations of state are also important to determine the composition of the compact matter and the thickness of the crust of a neutron star. The latter is important for neutrino generating (and hence neutrino emitting) processes and cooling mechanisms in a neutron star [Pe92].

In addition to this, the question of the maximum mass of neutron stars is important to determine the possibility of black holes in the galaxy. Examples of this can be Cyg X-1 and LMC X-3, which are candidates for galactic black holes [ST83, Øs89]. If the maximum mass of a neutron star becomes smaller, the probability for black holes formed after a supernova explosion naturally will be bigger.

The minimum mass of a stable neutron star is determined by setting the mean of the adiabatic index $\Gamma$, equal to the critical value for radial stability against collapse. The resulting minimum mass is $M \sim 0.1$ M$_\odot$ with a corresponding central density of $\rho \sim 10^{14}$ g cm$^{-3}$ and a radius $R \sim 200$ km. If we, however, include the possibility of quark stars, the possible minimal mass can take very low values, much lower than what we have for common neutron star models. The equilibrium configuration of the maximum mass is somewhat uncertain, but all microscopic calculations give us $M < 2{,}5$ M$_\odot$, and probably $M < 2{,}0$ M$_\odot$. Astronomical observations that give us neutron star parameters like total mass, radius and inertia are important, since these properties are sensitive to the models we choose for the microscopic structures of the star [ØØ91].

We can divide the density area within which an equation of state tax must work, into four different regions. The first region we have for $\rho \lesssim 10^4$ g cm$^{-3}$. This is the low density region, where matter consists of common cores, dominated by $^{56}$Fe.

The second region, we have for the area $10^4 \lesssim \rho \lesssim 4.3 \times 10^{11}$ g cm$^{-3}$. At around $\rho \sim 10^4$ g cm$^{-3}$ the density is large enough for the electrons to get the energy needed to break away from their respective cores, and form an electron gas. At $\rho \sim 10^7$ g cm$^{-3}$ the electrons become relativistic. As the density increases, we begin to get inverse beta decay, and the nuclei become more and more neutron rich. Neutrons continues to increase in number compared against protons until the neutron rich nuclei begin to "drip" out neutrons. This is the so-called "neutron drip" point, we talked about in Chapter 1. The free electrons are the source of an ever-increasing share of the total pressure while the density increases, and at the neutron drip the pressure is almost entirely due to electrons.

The third region is located in the area of $4{,}3 \times 10^{11} \lesssim \rho \lesssim 5 \times 10^{14}$ g cm$^{-3}$. The neutron drip-point,

$$\rho_d \approx 4,3 \times 10^{11} \text{ g cm}^{-3} \quad , \qquad (2.79)$$

is the density where the cores begin to dissolve and merge together. After this the matter consists of free electrons, neutrons and nuclei, until the nuclei become so neutron rich and eventually small, and the matter becomes so dense that the nuclei cease to exist as separate objects. This happens at a density near the core density,

$$\rho_n \approx 5 \times 10^{14} \text{ g cm}^{-3} \quad . \qquad (2.80)$$

In this area between $\rho_d$ and $\rho_n$, the matter will be composed of neutrons, protons and electrons. The nuclei disappear when we move into the upper part of this density range, since their binding energy decreases with increasing density. Just above $\rho_d$ the adiabatic index drops sharply. This is because the neutron gas with its low density contributes significantly to the density, but little to the pressure. The adiabatic index does not increase to more than 4/3 again until $\rho > 7 \times 10^{12}$ g cm$^{-3}$. This means that no stable stars can have central densities in this region [Øs89].

The nuclei are now becoming increasingly neutron rich as the density increases, and their stability decreases until the neutron fraction reaches a critical value. At this value the nuclei dissolve, mainly by fusing together. Since the nuclei at this point are very neutron rich, the matter in the nuclei is very similar to that of the free neutron gas outside. The neutron gas between the nuclei reduces the nuclei surface energy significantly, and it disappears when the matter inside nuclei becomes equal to the matter outside.

The fourth region is found in the area where $\rho > 5 \times 10^{14}$ g cm$^{-3}$. The properties of cold compact matter, and the related equations of state are reasonably well understood up to densities around $\rho_n$. For densities above $5 \times 10^{14}$ g cm$^{-3}$ the situation is not yet fully understood. We may here have a state with a complex mix of hyperons up to densities of about $10^{15}$g cm$^{-3}$, see for example Ref. [**Ca74**].

## 2.3 The Equation Of State

In the non-relativistic limit, we can assume that the nuclear forces are conservative and independent of the speed of the core. The force can thus be calculated from a static potential. The nucleon-nucleon potential must then be selected and incorporated into a many body calculation to obtain an equation of state. In the calculation of an equation of state there is an experimental condition that the selected interaction must provide us with the observed properties of nuclear matter at saturation. There are four parameters that must be reproduced: the density at the point where saturation occurs, the energy and compressibility of symmetric nuclear matter and the volume-symmetry coefficient which measures the curvature of the expression of the energy. It seems to be difficult to reproduce these parameters simultaneously with two-body potentials made only from scattering experiments in the laboratory. Good agreement between theory and experiment can however be hoped for by in addition introducing three-body interactions.

Relatively soft[1] equations of state have been proposed since the average system energy is attractive at nuclear densities. If you are using rigid equations of state, you can get potentials where the average interaction energy is dominated by the attractive parts of the potential at nuclear densities, but the repulsive part at higher densities. More rigid equations of state create important changes in the structure and the masses of heavy neutron stars. When the interaction energy becomes repulsiv above core densities, the pressure is better able to resist gravitational collapse. The result is that for stiff equations of state we will have a larger maximum mass. Neutron star models based on rigid equations of state will also have a lower central density, a larger radius and a thicker surface. Such differences are important when determining an upper limit to the mass of a neutron star, surface potential, inertia, precession frequencies, and so on.

For low densities $\rho < \rho_n$, where one expects the nuclear forces to be attractive, the pressure is somewhat softened when one includes interactions. For very high densities the equation of state, however becomes slightly stiffer due to the dominance of the repulsive core of the nuclear potential. This

---

[1]Relevant literature often refers to "soft" and "rigid" equations of state. A rigid equations of state will have higher pressure than a soft equations of state at a given temperature and density.

gives us
$$P \to \rho c^2, \text{ for } \rho = \varepsilon/c^2 \to \infty \;, \tag{2.81}$$
and the speed of sound approaches
$$c_s = (dP/d\rho)^{1/2} \to c \;. \tag{2.82}$$

At very high densities, above $10^{15}$ g cm$^{-3}$, it is estimated that the composition includes a substantial number of hyperons and nucleon interactions must be treated relativistically. Relativistical many body methods for strongly interacting matter is unfortunately not fully developed. The equations of state developed, also have many uncertainties about them. These uncertainties concerns problems such as the possibility of super-fluid neutrons and protons, for pion condensation, neutron solidification, possible phase transitions to quark matter and consequences of $\Delta$-nucleon resonance.

At densities considerably greater than $\rho_n$, it is no longer possible to describe the nuclear matter by using a non-relativistic many-body Schrödinger equation or by using a potential. The Meson-clouds surrounding the nucleons begin to overlap, and the image of specific, localizable, individual particles interacting through two-body forces, collapses. Even at a time before this collapse, various potentials that reproduce reliable low-energetic phase shift data, will give us other state equations. The potentials are here sensitive to the repulsive core region which is unaffected by the phase shift data. If the quarks are the fundamental building block of all strongly interacting particles, a description of nuclear matter at very high densities should involve them. When the nuclei are very close to each other, one would think that matter just above this density would go through a phase transition where quarks will begin to "drip" out of the nucleons. The result would then be quark matter, a degenerate Fermi liquid [Øs89].

# Chapter 3

# The composition of equations of state

## 3.1 Ideal Fermi gas of neutrons

A first approximation to the structure of a neutron star, we get by assuming a degenerate gas of non-interacting particles, an ideal Fermi gas. We will first look at the equations for such a gas made of neutrons. We thus assume a degenerate gas consisting of free neutrons. This is a very simple model of a neutron star, but it is still interesting to see how it behaves, and what distinguishes it from other subsequent calculations.

The particles in an ideal gas occupy energy states that can not be influenced by interactions between particles. The particles can be considered as waves, and wave properties can be used to determine the possible quantum states that are being occupied. If we assume that the particles are trapped in a box of volume $V$, the number of quantum states $g(p)$ will be given by [Ph99]

$$g(p) = g_s \frac{V}{h^3} 4\pi p^2 \quad , \tag{3.1}$$

where $h$ is Plancs constant, $p = \hbar k$ is the momentum of a free particle and $g_s$ is the number of independent polarizations. We will now work with neutrons, which, like electrons and protons are spin $\frac{1}{2}$ particles, with $g_s = 2$.

The internal kinetic energy of the gas depends on the density of the states, the energy of each quantum state and the number of particles in each state. The density of states, $g(p)\,dp$, is given by Equation (3.1). From special relativity we know that the energy $E$ of a neutron with mass $m_n$ in a quantum state with momentum $p$, is given by

$$E^2 = p^2c^2 + m_n^2c^4 \ . \tag{3.2}$$

The energy density of the gas, $\epsilon$, we can write as

$$\epsilon = \rho c^2 = \frac{1}{V}\int_0^\infty Ef(E)\,g(p)\,dp \ , \tag{3.3}$$

where $f(E)$ represents the average number of particles in a state of energy $E$. The particle density of the neutrons is given by

$$n_n = \frac{1}{V}\int_0^\infty f(E)\,g(p)\,dp \ . \tag{3.4}$$

Macroscopically the thermodynamic properties of the gas can be described by temperature $T$, the pressure $P$ and the chemical potential $\mu$. These parameters determine how the internal energy changes by the transfer of heat or entropy $S$, by compression or expansion, or by a transfer of particles. We express the thermodynamic relationships by

$$d\epsilon = V(TdS - PdV + \mu dN) \ . \tag{3.5}$$

In an ideal gas, we assume constant entropy and particle density. This simplifies the equation, and together with (3.3) we get for the pressure of an ideal gas:

$$\begin{aligned} P &= -\frac{1}{V}\frac{\partial \epsilon}{\partial V} \\ &= -\frac{1}{V}\int_0^\infty \frac{dE}{dV}f(E)\,g(p)\,dp \\ &= -\frac{1}{V}\int_0^\infty pv f(E)\,g(p)\,dp \ , \end{aligned} \tag{3.6}$$

when we have $dE/dV = -pv/3V$, where $v$ is the speed.

For an ideal gas of fermions the Pauli exclusion principle applies. No more than one particle can be in a given quantum state, and the average number $f(E)$ is

$$f(E) = \frac{1}{\exp[(E-\mu)/k_BT] + 1} \ , \tag{3.7}$$

where $k_B$ er the Boltzmann constant.

For completely degenerate fermions, where $T \to 0$ and thereby $\mu/kT \to \infty$, $\mu$ is the Fermi energy $E_F$, and

$$f(E) = \begin{cases} 1 & \text{for } E \leq E_F \\ 0 & \text{for } E > E_F \end{cases} . \tag{3.8}$$

We can now define the Fermi momentum by

$$E_F^2 = p_F^2 c^2 + m_n^2 c^4 . \tag{3.9}$$

If we now use the expression for $f(E)$ above, and the expression for $g(p)$ from Equation (3.1), we can use these to find the particle density $n_n$, the pressure $P$ and energy density $\epsilon$.

The particle density from Equation (3.4) then becomes

$$n_n = \int_0^{p_F} \frac{2}{h^3} 4\pi p^2 dp = \frac{8\pi}{3h^3} p_F^3 = \frac{1}{3\pi^2 \lambda_n^3} x^3 , \tag{3.10}$$

where $\lambda = \hbar/m_n c$ is the Compton wavelength of the neutron. We have also introduced a dimensionless Fermi momentum $x$, which is given by

$$x = \frac{p_F}{m_n c^2} . \tag{3.11}$$

The pressure in Equation (3.6) can be written as

$$P = \frac{m_n c^2}{\lambda_n^3} \phi(x) , \tag{3.12}$$

where

$$\phi(x) = \frac{1}{8\pi^2}(x(1+x^2)^{1/2}\left(\frac{2x^2}{3} - 1\right) + \ln[x + (1+x^2)^{1/2}]) . \tag{3.13}$$

Similarly, the energy density of (3.3) can be expressed by

$$\epsilon = \rho c^2 = \frac{m_n c^2}{\lambda_n^3} \chi(x) , \tag{3.14}$$

where

$$\chi(x) = \frac{1}{8\pi^2}(x(1+x^2)^{1/2}(1+2x^2) - \ln[x + (1+x^2)^{1/2}]) . \tag{3.15}$$

We can also express the Fermi-momentum by the particle density by help of Equation (3.10),

$$p_F = \left(\frac{3n_n}{8\pi}\right)^{1/3} h . \tag{3.16}$$

27

One can easily show that when gas particles are non-relativistic or ultra-relativistic, the pressure is directly proportional to the kinetic energy density of the gas. At the same time, one can show that the function $\phi(x)$ in Equation (3.12) can be simplified, so that we get a simple expression for the pressure as a function of particle density.

For non-relativistic neutrons, $p_F \ll m_n c$, we will have $\epsilon = m_n c^2 + p^2/2m$ and $v = p/m$, and we will get the following expression for the pressure:

$$P = \frac{2N}{3V}\left\langle \frac{p^2}{2m} \right\rangle = \frac{2}{3} \text{ of the kinetic energy density.} \qquad (3.17)$$

Simultaneously, simplified from (3.12) we get,

$$P = \frac{h}{5m_n}\left[\frac{3}{8\pi}\right]^{2/3} \cdot n_n^{5/3}. \qquad (3.18)$$

For ultra-relativistic neutrons, $p_F \gg m_n c$, we will get $\epsilon = pc$ and $v = c$, and we will get for the pressure:

$$P = \frac{N}{3V}\langle pc \rangle = \frac{1}{3} \text{ of the kinetic energy density.} \qquad (3.19)$$

At the same time, simplified from (3.12), we get

$$P = \frac{hc}{4}\left[\frac{3}{8\pi}\right]^{1/3} \cdot n_n^{4/3}. \qquad (3.20)$$

## 3.2 Quark matter

When nuclear matter is compressed to such high densities that the nucleon cores overlap considerably, it is expected that the nucleons merge and undergo a phase transition to quark matter. We therefore expect a transition from hadronic degrees of freedom to quark-degrees of freedom at high densities. The matter would then undergo a phase transition where quarks would begin to drip out. The result would be a degenerate Fermi liquid of quark matter, but to calculate the effects of this transition in neutron stars, we must know the equations of state for both quark matter and hadronic matter.

Since no free quarks have been observed, it is believed that they are kept inside hadrons by a force that increases if we try to separate the quarks. According to quantum chromodynamics this power will nevertheless be correspondingly weaker when the quarks are squeezed together. Therefore one has suggested that quarrk matter, as an approximation at sufficiently high densities, can be treated to a first order as an ideal relativistic Fermi gas. This will provide a relatively soft equation of state. In an ideal Fermi gas model of quark matter, we assume a system of non-interacting quarks moving freely. At densities corresponding to normal nuclear matter, $\rho = 2,3 \cdot 10^{14} \text{g/cm}^3$, quarks are known to interact very strongly with each other, so at these densities this model will fit poorly. When the quarks on the other hand are very close to each other, we can assume that they move freely, as if they were not bound at all. Such an ideal Fermi gas model of quarks can thus be useful at very high densities.

For the final densities, one should take into account quark interactions. At moderately high densities, one can use a perturbational expansion in the coupling constant of the strong interaction, and assume that the quarks are asymptotically free. At lower densities the quarks can however occur bound, and similar phenomenological models have then been developed. One such model is the "MIT bag" model, where the quarks in the nucleons have their freedom of movement restricted within a certain spatial area, a "bag". The volume of this bag is kept finite by a pressure $B > 0$, called the bag constant. $B$ here is the energy density needed to "inflate" the bag. The total energy density of the quark matter will be the non-interacting Fermi gas contribution plus $B$ [**HH00**].

By comparing the equations of state and calculated the energy curves, we can then estimate a possible phase transition from neutron matter to quark matter. We must then see whether this phase transition may occur at densities below the maximum density of a stable neutron star.

## 3.3 Superfluid baryon matter

The presence of super-fluid neutrons in the crust and the interior of neutron stars, is seen as well-established facts in the physics of compact stars. One assumes two different layers of super-liquid matter. In the outer core, where we have relatively lower densities, we have a range of super fluid neutrons. When the density increases inwards in the star, the cores are dissolved from the crust, and one assumes an area consisting of a quantum fluid of neutrons and protons in beta equilibrium. When the core of the star then is reached, we assume that these super liquid phases disappear [**HH00**].

## 3.4 Kaon condensation

The idea that the ground state of baryon matter may contain a Bose-Einstein condensate of kaons comes from Kaplan and Nelson [**KN86**], and has since been discussed in many publications. Because of the attraction between $K^-$ and the nucleons, the energy of the kaon will decrease with increasing density until it falls below the chemical potential of the electron in the neutron star matter in $\beta$-equilibrium. When this happens, we get a Bose-Einstein condensate of $K^-$-kaons. Depending on the parameters used, particularly the strangeness of the proton, one finds that $K^-$ condensates at densities above $\sim$ 3-4$\rho_0$, where $\rho_0 = 0.16$ fm$^{-3}$ is the equilibrium density of nuclear matter. This makes the equation of state softer, and lowers the maximum mass of the neutron star [**PBP97, HH00**]. One can then compare this with the central density of a neutron star with mass 1.4 M$_\odot$, which is at about $\sim$ 4$\rho_0$.

In neutron matter at low densities, the distance between the particles is much larger than the range of the interactions, and the kaons will then interact many times with the same nucleon before it encounters and interacts with another nucleon. We can then use the diffusion length as the effective kaon-neutron interaction, $a_{K-N} \simeq -0,41$ fm, where we here ignore the small proton fraction we find in the nuclear matter. The kaon energy differs from the rest mass by the Lenz potential, the optical potential is achieved in the momentum approximation bevegelsesmengdeapproksimasjonen [**HH00**],

$$\omega_{Lenz} = m_K + \frac{2\pi}{m_R} a_{K-N} n_{NM} \quad . \qquad (3.21)$$

If the hadron masses decrease even more, condensation will occur at lower densities.

In neutron matter at high densities, the distance between the particles is much smaller than the range of interactions, and the kaon will interact with many nucleons within this distance. The kaon will then be exposed to the field from many nucleons and the kaon energy will differ from the rest mass of the Hartree potential [**HH00**]:

$$\omega_{Hartree} = m_K + n_{NM} \int V_{K-N}(r) \, d^3r \quad . \qquad (3.22)$$

The Hartree potential is considerably less attractive than the Lenz potential, and already at densities where the particle spacing is comparable to the range of the kaon-nucleon ($KN$)-interaction, the kaon-nucleon and nucleon-nucleon-correlations begin to reduce the attraction in the $K^-N$-interaction significantly. These correlations begin to have a significant effect from the point where their range is comparable to the distance between the nucleons. The transition from the Lenz potential at low densities to the Hartree potential at high densities, appears already well below the nuclear density. For measured $K^-N$-diffusion lengths and reasonable ranges of interactions, the attraction is reduced by a factor of about 2-3 in the core of neutron stars. Relativistic effects reduce the additional attraction at high densities. The result is then that a kaon condensate becomes less likely in neutron stars due to nuclear correlations, but if the kaon masses decrease with the density, condensation will start at lower densities [**HH00**].

## 3.5 Pion condensation

Just like kaon condensation, pion condensation is possible in neutron star matter at high densities. If we neglect the effect of strong correlations between pions and the matter when we modify the self-energy of the pion, we find that it becomes energetically favorable for a neutron at the top of the Fermi sea to transform to a proton and $\pi^-$ when

$$\mu_n - \mu_p = \mu_e > m_\pi , \qquad (3.23)$$

where $m_\pi = 139.6$ MeV is the rest mass of the $\pi^-$-pion. At the saturation density of nuclear matter the chemical potential of the electron is ~100 MeV, and one can therefore expect a condensation of $\pi^-$ at a slightly higher density. One can not, however, neglect interactions between the pion and the background matter. Such interactions can increase the self-energy of the pion, and thus increase the threshold density. Depending on the chosen parameters, the critical density for pion condensation can vary from $n_0$ to $4n_0$. These are questions yet to be solved in a satisfactory manner. Models with strong correlations between nucleons suppress condensation, both via $\pi NN$ and $\pi\Lambda N$ so that pion condensation in neutron star matter may not occur. On the other hand, in addition to the pion condensate with the charge, we may also obtain a $\pi^0$-condensate through the reaction

$$n \to n + \pi^0 , \qquad (3.24)$$

if the effective mass of the $\pi^0$ in the medium is zero [**HH00**]. There is however, still some uncertainty about the $\pi^0$-condensation.

The existence of pion condensates will be able to accelerate the cooling of neutron stars dramatically. If it exists, we will be able to get quasi-particle $\beta$-decays via

$$N \to N' + e^- + \bar{\nu}_e \qquad (3.25)$$

and the corresponding inverse reaction. Here the quasi-particles $N$ and $N'$ are linear combinations of neutron and proton states in the pion sea. The pion condensates allow both energy and momentum to be conserved in the reaction, which is analogous to the direct Urca process. A simplified version of this reaction has also been proposed, where one looks at cooling via decay of free pions [**ST83**].

## 3.6 Phase Transitions

We can imagine many different types of phase transitions in a neutron star. We can, for example, have a transition from a nuclear liquid to gas in the inner crust, and in the interior, we can have the presence of quark matter and/or condensates of kaons, pions, hyperons and so on.

Normally the transition between two phases occurs at a certain pressure, a certain temperature and a given chemical potential. As a consequence, one expects that the density jumps discontinuously on the border between the two phases. This is true not only for the one-component systems, such as water that freezes or evaporates, but also for some two-component systems, such as neutral nuclear matter in $\beta$-equilibrium. Electrical neutrality requires that the number density of electrons and protons to be equal,

$$n_p = n_e \ . \tag{3.26}$$

$\beta$−equilibrium requires that the difference in the chemical potential of neutrons and protons, corresponds to the chemical potential of the electron in nuclear matter (NM = nuclear matter),

$$\mu_n = \mu_p + \mu_e^{\text{NM}} \ . \tag{3.27}$$

These two conditions determine two out of three components, leaving only one independent variable, the baryon density.

In quark matter we have a similar requirement of charge neutrality

$$\frac{2}{3}n_u - \frac{1}{3}n_d - \frac{1}{3}n_s = n_e \ . \tag{3.28}$$

Similarly the $\beta$−equilibrium requires (QM = quark matter),

$$\mu_d = \mu_s = \mu_u + \mu_e^{\text{QM}} \ . \tag{3.29}$$

Over the past few decades, many researchers have looked at the properties of neutron stars with a core of quark matter. In such hybrid stars it is estimated that each of the two phases are separatley electrically neutral, as in Equation

(3.26) and (3.28), and in $\beta$−equilibrium separatley, as in (3.27) and (3.29). Gibbs conditions, $P_{NM} = P_{QM}$ and $\mu_n^{NM} = \mu_n^{QM}$, gives us a certain density, where the two neutral phases can coexist.

When we consider the distinct phases of matter, such as pure baryon or quark matter, the composition of matter will thus be determined by the conditions set by the chemical and electrical equilibrium conditions.

# Chapter 4

# Hyperon matter

## 4.1 Description of compact matter with hyperons

We will now seek to give a brief introduction to the consequences we get to the equation of state from hyperons, and which interactions we must take into account in order to provide an adequate description of the densities and equations of state of a neutron star with hyperons.

Much of the densities inside neutron stars can be expressed satisfactorily by means of the degrees of freedom of the nucleons. This includes the areas from the interior of the crust to the outer parts of the core - areas that range from 0.5 to 2-3 times the saturation density of nuclear matter. There are lots of experimental and theoretical data that support the assumption that nucleons do not lose their individuality in such a compact matter, see for example [**MST90**]. This means that the properties of nucleons at these densities is similar to those of free nucleons. The density range above gives us distances between nucleons of the order of magnitude ~1 fm. At such distances, there is little overlap between the different nucleons. We therefore assume that they still have a behavior similar to individual nucleons, and that the effects of the overlap we have, can be incorporated in the description of a two-nucleon interaction.

To briefly outline the nucleon-nucleon-interactions in a nuclear medium, one can then use a as simple many body system as possible, namely the Brueckner-Hartree-Fock (BHF) method, which we termed as the lowest order Brueckner theory (LOB) [**HH00**].

Following the conventional many body approach, we split the Hamiltonian in the following manner, $H = T + V$, where $T$ is the kinetic energy and $V$ is the free nucleon-nucleon (NN)-interaction, into a uperturbed part $H_0 = T + U$, and an interacting Part $H_1 = V - U$, such that

$$H = T + V = H_0 + H_1 \ . \tag{4.1}$$

We have here introduced a single-particle help potential $U$. If we choose $U$ so that $H_1$ becomes small, we can assume that perturbative many body theory can be applied. An obstacle for such a perturbative treatment is the fact that the free NN interaction gives us matrix elements who are very large, or diverge at short distances between the cores, making a perturbative approach almost impossible. To solve this problem, we introduce the reaction matrix $G$, which is given by the solution of the Bethe-Goldstone equation in operator form

$$G(\omega) = V + VQ\left[1/(\omega - QH_0Q)\right]QG \ . \tag{4.2}$$

In a partial wave represenattion, this becomes

$$G_{ll'}^{\alpha T_z}(kk'K\omega) = V_{ll'}^{\alpha T_z}(kk') + \sum_{l''} \int \frac{\mathrm{d}^3 q}{(2\pi)^3} V_{ll''}^{\alpha T_z}(kq) \frac{Q^{T_z}(q,K)}{\omega - H_0} G_{l''l'}^{\alpha T_z}(qk'K\omega), \tag{4.3}$$

where the variable $K$ is the momentum corresponding to the movement of the mass center. This equation allows us to calculate the energy per particle, and thus also the corresponding proton fraction by setting equilibrium conditions and including muons. We can thus obtain knowledge of the composition of, and which phases are present in, the compact matter [**HH00**].

$\mathcal{E}$, the non-relativistic energy per nucleon, is given by

$$\mathcal{E} = \frac{1}{A} \sum_{h \leq k_F} \frac{k_h^2}{2m} + \frac{1}{2A} \sum_{h \leq k_F, h' \leq k_F} \langle hh' | G(E = \varepsilon_h + \varepsilon_{h'}) | hh' \rangle_{AS} \ . \tag{4.4}$$

In this equation, we have suppressed the isospin indices for the Fermi momenta. The equation above is calculated for various proton fractions $x_p$, and thus becomes a function both of density $n$ and $x_p$. Different interactions as the charge-dependent Bonn interaction (CD-Bonn) [**MSS96**], the three Nijjmegen-potentials, Nijm-I, Nijm-II and Reid93 [**SKT94**] and Argonne $V_{18}$ [**WSS95**] provides various functions for the total potential energy per nucleon. These new nucleon-nucleon-interactions nevertheless give us relatively similar energies per particle in neutron star matter.

We will now briefly outline how more sophisticated many-body calculations on the two-body level can provide similar results for neutron matter. This is because the contribution from three-body interactions possibly may change this picture somewhat.

The BHF method in its simplest form, does thus not provide an adequate description of compact matter. When one reaches densities above the saturation density $n_0$ it is estimated that more complex many-body terms are significant. Such contributions originate in things like core polarization effects, effective three-body and many-body diagrams, and eventually the inclusion of three-body forces is assumed to be important at densities above $n_0$. Possible improvements to the LOB theory may be the summation to infinite order of the sequence of particle-particle hole-hole diagrams (PPHH), the three-hole line results of Baldo et al [**SBG98**], or calculations with three-body powers of Akmal et al [**APR98**].

The summation of PPHH diagrams means that the Pauli operator is expanded to prevent the spread of the hole-hole (HH) intermediary states. In addition to summing particle-particle (PP) intermediary states up to infinite order, we must now also sum one larger class of diagrams containing HH intermediary states. One of the thoughts behind the PPHH diagram method for nuclear matter, is thus to include charts with HH-correlations to infinite order.

Baldo et al. [**SBG97, SBG98**] calculates the sum of so-called three hole-line diagrams. The entire set of such three hole-line diagrams can be divided into three main groups, and each of these three hole-line contributions are quite large. There is still a large degree of cancellation between different addents, and the total three hole-line contribution is therefore significantly less than the two-body contribution. The interesting thing with the calculations of Baldo et al. [**SBG98**], is that they lead to results for the three hole-lines, which is close to the total performance of two hole-lines, PPHH diagrams not included. Despite these methodological advances, there are still several classes of diagrams to be summarized. It can be proved that there is a minimum set of diagrams that must be added to make the physics of these many body systems correct [**HH90**]. This is the so-called Parquet class.

Since this class is difficult to summarize, one has made calculations of many body clusters, and it is through this that we have the results of Akmal, Pandharipande and Rave Hall [**APR98**]. We will not go into this in detail, the interested reader can find these calculations, as well as the discussions in [**APR98, HH90**].

In summary, the inclusion of phenomenological three-body forces in non-relativistic calculations is necessary to improve the saturation properties of a microscopic estimated equation of state. Inclusion of other many body effects at low densities gives results similar to those from LOB theory. Even with three-body forces, one still does not get a completely accurate expression of the binding energy. We must therefore consider further corrections from relativistic effects.

The properties of neutron stars depends on densities that are of an order of magnitude higher than the one observed for ordinary nuclei. At such densities we will encounter relativistic effects. Relativistic approaches, such as the Dirac-Hartree and the Dirac-Hartree-Fock approach, has, among other things, managed to quantitatively reproduce spin observables that has been poorly described in non-relativistic theory. Introduction of a highly attractive scalar component and a repulsive vector component in the self-energy of the nucleon has been very important for these methods. Furthermore, a relativistic extension of Brueckner theory has been developed. This so-called Dirac-Brueckner theory gives us self-consistent expressions for the relativistic single-particle energies and wave functions. Dirac-Brueckner-theory is based on nucleon-nucleon-interactions, on meson exchange and is a non renormalizable theory [**HH00**]. Here, one uses media modified quantities to find a relativistic $G$-matrix, and its single-particle potential.

$$u_i = \sum_{h \leq k_F} \frac{\tilde{m}_i \tilde{m}_h}{\tilde{E}_i \tilde{E}_h} \langle ih | \tilde{G}(\tilde{E} = \tilde{E}_i + \tilde{E}_h) | ih \rangle_{AS} \quad , \tag{4.5}$$

where $\tilde{m} = m + U_S$ and $\tilde{E}_i = \tilde{E}(p_i) = \sqrt{\tilde{m}_i^2 + \mathbf{p}_i^2}$ are media modified quantities for matter and energy, and $\tilde{G}$ is the relativistic $\tilde{G}$-matrix. By iterative methods one can then find a relativistic equation for the energy per nucleon, $\mathcal{E}/A$.

$$\begin{aligned}\mathcal{E}/A &= \frac{1}{A} \sum_{h \leq k_F} \frac{\tilde{m}_h m + k_h^2}{\tilde{E}_h} + \\ &\quad \frac{1}{2A} \sum_{h \leq k_F, h' \leq k_F} \frac{\tilde{m}_h \tilde{m}_{h'}}{\tilde{E}_h \tilde{E}_{h'}} \langle hh' | \tilde{G}(\tilde{E} = \tilde{E}_h + \tilde{E}_{h'}) | hh' \rangle_{AS} - m \quad .\end{aligned} \tag{4.6}$$

An alternative approach is to include the relativistic boost corrections to the non-relativistic NN-interactions. The NN-scattering data are reduced to the center of mass system, and customized using phase shifts calculated from the NN-interaction in the current frame of reference. The interaction you get through this method describes the NN-interaction in the center of mass system, where the total momentum, $\mathbf{P}_{ij} = \mathbf{p}_i + \mathbf{p}_j$, is zero.

Calculations of such effects allows us to develop more accurate expressions of the single-particle energy and the proton fractions, and it gives us better insight into the composition of the compact matter in neutron stars. In the next section we will look at the equation of state of Akmal et al [**APR98**] with boost corrections and three-body forces.

## 4.2 Parameterization of the equation of state for nuclear matter

Since we assume that three-body forces are important, we will apply the equation of state of Akmal et al [**APR98**] in our discussion of the complex phase and structure of a neutron star. We here prefer a non-relativistic equation of state with boost corrections. Up to a few times the saturation density, $n_0 = 0.16$ fm$^{-3}$, we know the equation of state for nuclear matter rather well. Detailed knowledge of the equation of state is essential to be able to say anything about the existence of e.g.. pion condensates or structures inside the neutron stars [**HH00**]. For the more global properties of neutron stars, we still adopt a simple form of binding energy per nucleon in nuclear matter. This consists of a compression term and a symmetry term

$$\mathcal{E} = E_{comp}(n) + S(n)(1-2x_p)^2 = \mathcal{E}_0 u \frac{u - 2 - \delta}{1 + \delta u} + S_0 u^\gamma (1-2x_p)^2 \ . \quad (4.7)$$

Here $u = n/n_0$ is the relationship between the baryon density and the saturation density, and we have defined the proton fraction as $x_p = n_p/n$. The compression term in Equation (4.7) is parameterized in a simple form that reproduces the saturation density, the binding energy and compressibility. The binding energy without Coulomb energies at saturation density is $\mathcal{E}_0 = -15.8$ MeV. The parameter $\delta = 0.2$ is determined by adapting the energy per nucleon at high densities to the equation of state of Akmal et al [**APR98**] with three-body forces and boost corrections, where one uses the corrected values from Table 6 in [**APR98**]. The corresponding compressibility is $K_0 = 18 \mathcal{E}_0/(1+\delta) \simeq 200$ MeV, which is consistent with the experimental value. For the symmetry term, we get $S_0 = 32$ MeV and $\gamma = 0.6$ for the best fit. In Figure 4.1 we compare the parameterization with the results of Akmal et al [**APR98**]. As we see the results match quite well except at very high densities. At these densities the equation of state of Akmal et al [**APR98**] nevertheless becomes superluminal, and therefor must be wrong. Akmal et al [**APR98**] provides a much more sophisticated adaptation, but it's still amazing that results from such a simple quadratic equation can correspond so well with data from calculations on the microscopic level. When we look at the uncertainties of equations of state at high densities, one must be able to say that this parameterization is within the current margins of error [**HH00**]. The correlation between the microscopic calculations of Akmal et al [**APR98**] and the simple parameterization in Equation (4.7) may imply that the central many body physics near the saturation density comes only from two-and three body terms in $\mathcal{E}$. The reason for this is [1] that three body terms are proportional to $n^3$, while two body terms are proportional to $n^2$. With three body terms we mean both effective interactions, and contributions from real three body forces. The calculation of these real powers still is an unsolved problem [**HH00**].

If we now restrict our attention to matter with electrons only, we can easily obtain an analytical expression for the proton fraction using the asymmetry parameter $x$. We then have the equilibrium states of matter in $\beta$-equilibrium in Equation (4.8)-(4.9),

$$\mu_n = \mu_p + \mu_e \; , \tag{4.8}$$

and

$$n_p = n_e \; , \tag{4.9}$$

where $\mu_i$ and $n_i$ are the chemical potential and the number density in fm$^{-3}$ of particle type $i$. When one thwn uses the definition of the chemical potential

$$\mu_i = \left(\frac{\partial \varepsilon}{\partial n_i}\right) \; , \tag{4.10}$$

one finds that

$$\mu_e = \frac{1}{n}\frac{\partial \varepsilon}{\partial x_p} \; . \tag{4.11}$$

When one then assumes ultra relativistic electrons, we easily find the proton fraction given by

$$nx_p = \frac{(4S_0 u^\gamma (1-2x_p))^3}{3\pi^2} \; . \tag{4.12}$$

When we define

$$a = \frac{2(4S_0 u^\gamma)^3}{\pi^2 n} \; . \tag{4.13}$$

Equation (4.12) is reduced to

$$3x^3 + ax - a = 0 \; , \tag{4.14}$$

where $x = 1 - 2x_p$. Since we are always looking at solutions for densities above zero, the cubic equation for $x$ will have an analytical solution that is real and given by

$$x = -\frac{2\sqrt{a}}{\tan(2\psi)} \; , \tag{4.15}$$

---

[1]This argument is for the energy density, $\varepsilon = \mathcal{E}n$.

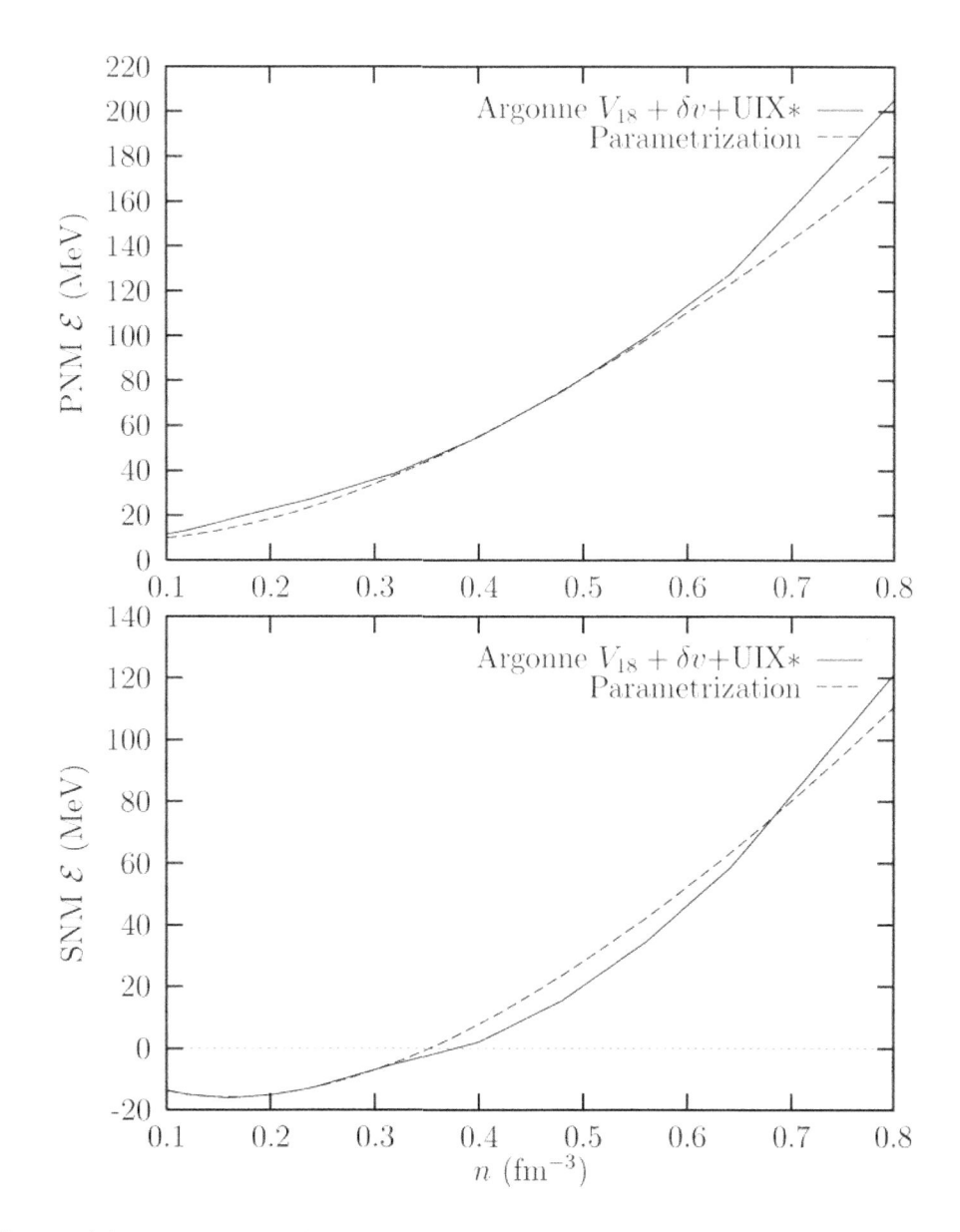

Figure 4.1: Top panel: Comparison of the parametrized equation of state from Equation (4.7) and the results of Akmal et al [APR98] with boost corrections and three body forces for pure neutron matter. Bottom panel: The corresponding results for standard nuclear matter.

with $\tan\psi = \left(\tan\frac{\phi}{2}\right)^{\frac{1}{3}}$ and $\tan\phi = -2\sqrt{a}/3$. Note that $x$ only depends on the total baryon density $n$. This means that our parameterization of the equation of state now can be rewritten for matter in $\beta$-equilibrium as

$$\mathcal{E} = \mathcal{E}_0 u \frac{u - 2 - \delta}{1 + \delta u} + S_0 u^\gamma \left(\frac{2\sqrt{a}}{\tan(2\psi)}\right)^2 , \qquad (4.16)$$

where $\mathcal{E}$ is an analytical function of the density only.

In Figure 4.2 we see the qualitative results for other observables than the energy per particle for proton fractions from the simple expressions in Equation (4.15) for this approximation to the equation of state of Akmal et al [**APR98**].

In the same figure we also show the resulting energy per nucleon in matter in $\beta$-equilibrium, and compare it with the results of Akmal et al [**APR98**] for different values of $\delta$. Note that the proton fraction does not depend on the value of $\delta$, see Equation (4.12). As we thus can see, the equation of state with $\delta = 0$ is the stiffest, and thus gives a superluminal behavior at densities greater than $n \approx 1.0$ fm$^{-3}$. The shape of (4.7), with $\delta$, gives us a smooth extrapolation of small and large densities with a correct behavior in both limits. The binding energy per nucleon $E/A = \mathcal{E}$ is linear with respect to the number density. In the dilute limit, this is the optical Lenz potential. At high densities linearity is required by the condition that the sound speed $v_s^2 = \partial P/\partial \varepsilon$ must not exceed the speed of light. This justifies the introduction of the parameter $\delta$ in the parameterization and explains the deviation from the results of Akmal et al [**APR98**] at densities greater than $0.6 \sim 0.7$ fm$^{-3}$, see Figures 4.1 and 4.2.

## 4.3 Hyperon matter

At the core density the chemical potential of the electron is ~110 MeV. As soon as you have exceeded the rest mass of the muon, it becomes energetically favorable for an electron at the top of Fermi surface to decay to a $\mu^-$. In this way we get a lake of degenerate negative muons.

Similarly, the energetic neutrons decay to $\Lambda$-hyperons, when the chemical potential of neutrons becomes sufficiently large. This occurs through weak interactions that do not preserve the strangeness, which leads to a $\Lambda$-Fermi lake with $\mu_\Lambda = \mu_n$. If we neglect the interactions, or assume that their effect is small, we till assume $\Sigma^-$ to appear at lower densities than the $\Lambda$, via

$$e^- + n \rightarrow \Sigma^- + \nu_e , \qquad (4.17)$$

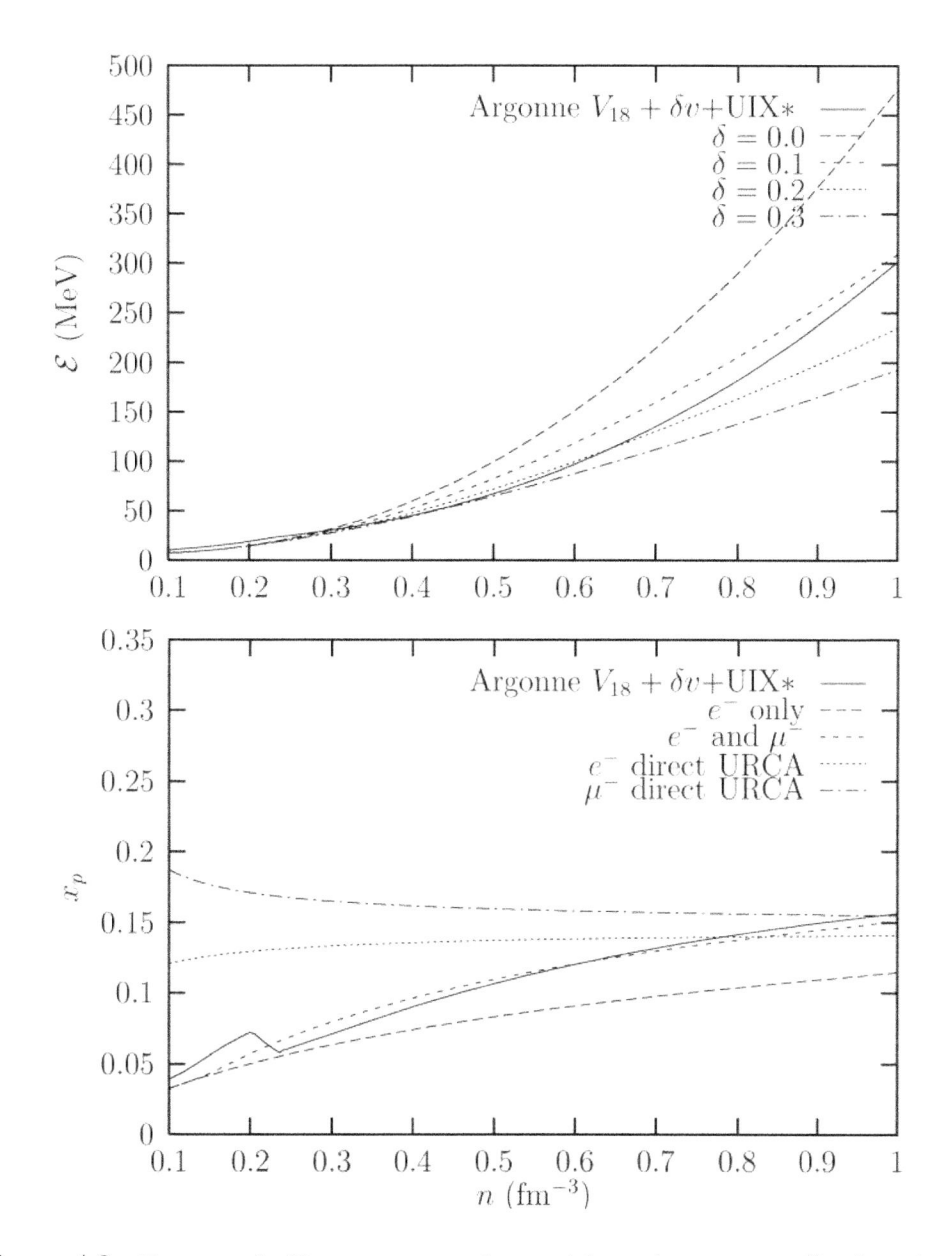

Figure 4.2: Top panel: Energy per nucleon without lepton contributions in $\beta$-stable matter for the parametrized equation of state in Equation (4.7>) for $\delta = 0, 0.13, 0.2, 0.3$ and the results of Akmal et al [APR98] with boost corrections and three body forces. Bottom panel: The corresponding proton fraction $x_p$.

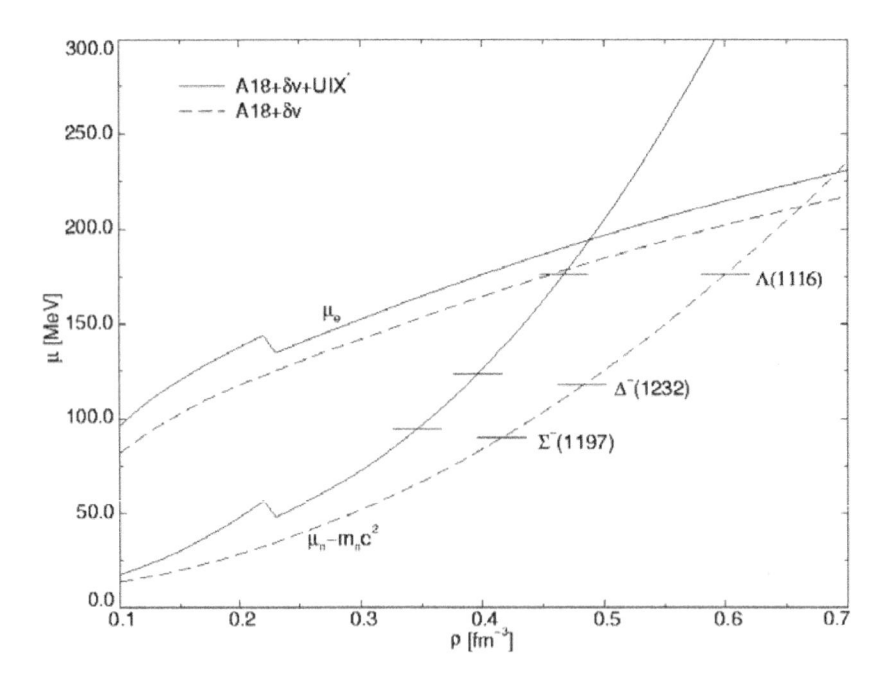

Figure 4.3: The chemical potential for neutrons and electrons in matter in $\beta$-equilibrium for the models $V_{18} + \delta v +$UIX$^*$ (continuous line) and $V_{18} + \delta v$ (dashed line). The treshold densities for when the non-interacting hyperons pop up is marked by horizontal line segments. From Reference [APR98].

even though $\Sigma^-$ is more massive. The reason for this is that the above process removes both an energetic neutron, and an energetic electron, while the disintegrations of a $\Lambda$, which is neutral, removes only the neutron. The negatively charged hyperons occur in the ground state of matter when their masses equal $\mu_e + \mu_n$, while the neutral $\Lambda$ hyperonet first occurs when $\mu_n$ is equivalent to its mass. Since the chemical potential of the electron is greater than massedifferensen $m_{\Sigma^-} - m_\Lambda = 81.76$ MeV, $\Sigma^-$ will occur at lower densities than the $\Lambda$. We show this in Figure 4.3 where we plot the chemical potentials for electrons and neutrons in matter in $\beta$-equilibrium. The treshold densities for $\Sigma^-$, $\Lambda$ and the isobar $\Delta^-$ is indicated by the horizontal lines.

Since we here focus on the many body approaches, we will start with the parameterization of the free baryon-baryon-potentials for the complete baryon octet defined by Stokes and Rijk in [**SL99**] and further used in [**HH00**]. This potential model seeks to describe all interaction channels with strangeness from $S = 0$ to $S = -4$, and it is based on the SU(3) developments from the Nijmegen potential models [**RSY98**] for the $S = 0$ and $S = -1$ channels. These are adapted to the available experimental data, and put constraints on the free parameters in the model. In our discussion we use the interaction

NSC97e version of Reference [**SR99**], since this does not change the results for matter in $\beta$-equilibrium significantly.

The next step introduces the effects of the nuclear medium. We construct the $G$-matrix, which takes short-range correlations into the calculation for all strangeness, and solves the equation for the single-particle energies of the different baryons. The $G$-matrix is given by

$$\langle B_1 B_2 | G(\omega) | B_3 B_4 \rangle = \langle B_1 B_2 | V | B_3 B_4 \rangle + \sum_{B_5 B_6} \langle B_1 B_2 | V | B_5 B_6 \rangle \frac{1}{\omega - \varepsilon_{B_5} - \varepsilon_{B_6} + i\eta} \times \langle B_5 B_6 | G(\omega) | B_3 B_4 \rangle \quad . \quad (4.18)$$

Here $B_i$ represents all possible baryons $n$, $p$, $\Lambda$, $\Sigma^-$, $\Sigma^0$, $\Sigma^+$, $\Xi^-$ and $\Xi^0$ and that the quantum numbers like spin isospin, strangeness, momentum and angular momentum. Intermediate states $B_5 B_6$ are those allowed by the Pauli principle and the energy variable $\omega$ is the starting energy defined by the single-particle energies of the the incoming external particles $B_3 B_4$. The $G$-matrix is solved by using relative and center of mass coordinates, see for example [**SL99**] for details. The single-particle energies are given by

$$\varepsilon_{B_i} = t_{B_i} + u_{B_i} + m_{B_i} \quad , \quad (4.19)$$

where $t_{B_i}$ is the kinetic energy, and $m_{B_i}$ is the mass of baryon $B_i$. The single-particle potential $u_{B_i}$ is defined by

$$u_{B_i} = \text{Re} \sum_{B_j \leq F_j} \langle B_i B_j | G(\omega = \varepsilon_{B_j} + \varepsilon_{B_i}) | B_i B_j \rangle \quad . \quad (4.20)$$

The momentum of the intermediate state $B_j$ is limited by the size of Fermi surface $F_j$ for particle type $B_j$. The final equation is in the shape of Goldstone diagrams in Figure 4.4. Diagram (a) represents contributions from nucleons only as hole states, while diagram (b) only has hyperoner as holes states in the case we have a final hyperon fraction in neutron star matter in $\beta$-equilibrium.

To satisfy the equations for matter in $\beta$-equilibrium,

$$\begin{aligned} \mu_{\Sigma^-} &= \mu_{\Delta^-} = \mu_n + \mu_e \ , \\ \mu_\Lambda &= \mu_{\Sigma^0} = \mu_{\Delta^0} = \mu_n \ , \\ \mu_{\Sigma^+} &= \mu_{\Delta^+} = \mu_p = \mu_n - \mu_e \ , \\ \mu_{\Delta^{++}} &= \mu_n - 2\mu_e \ , \end{aligned} \quad (4.21)$$

we must solve the Equations (4.18) and (4.19) to find the single-particle energies of particles involved in the corresponding Fermi momentums. For

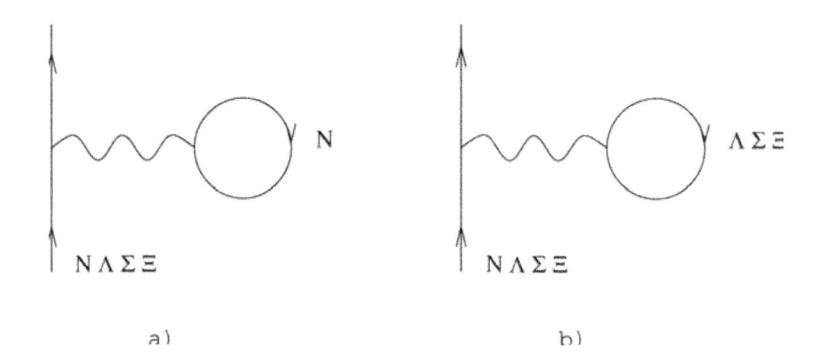

Figure 4.4: Goldstone diagram for the single-particle potential $u$. a) represents the contribution from nucleons only as hole states while b) includes only hyperons as hole states. The wavy line represents the $G$-matrix.

any total baryon densitytetthet $n = n_N + n_Y$, the density of nucleons plus hyperons, the Equations (4.18) and (4.19) are solved for five nucleon fractions and five hyperon fractions. For each nucleon and hyperon fraction, three proton fractions and three fractions for relevant hyperons are calculated. The set of equations in Equation (4.21) is the solved by interpolation between different nucleon and hyperon fractiones [**HH00**].

The many body procedure outlined above, is the Brueckner-Hartree-Fock (BHF) method to lowest order, expanded to the hyperon sector. This means that we only look at two-body interactions. It is well known from studies of nuclear matter and neutron star matter with only nucleonic degrees of freedom, that three-body interactions are important to reproduce the saturation properties of nuclear matter, see [**APR98, HH00**]. To incorporate such effects, we replace the contribution of the self-energies of protons and neutrons that comes only from the internediate states (see diagram (a) of Figure 4.4) with the one from [**APR98**]. Here, the Argonne $V_{18}$-nucleon-nucleon interaction is used with relativistic boost corrections and an adapted three-body interaction model. We use the parameterization of Equation (4.7) with $\delta = 0.2$. In the discussion below, we will present two sets of results for matter in $\beta$-equilibrium, one where the nucleon contributions to the self-energy are obtained from the baryon-baryon-potential model of Stokes and Rijk [**SR99**], and one where nucleon contributions are replaced with the results from [**APR98**] in the parameterization discussed in Equation (4.7). In the discussion in this section, we will note these results with the APR98. All the hyperon contributions are still calculated with the baryon-baryon-interactions of Stokes and Rijk [**SR99**].

These models for the pure nucleon part combined with the hyperon contribution provides us with the composition of matter in $\beta$-equilibrium, up to the total number density of the baryons $n = 1.2$ fm$^{-3}$, as shown in Figure 4.5. The corresponding energies per baryon is shown in Figure 4.6 for both pure nucleon matter (BHF and APR98 with pure nucleon matter) and hyperon matter (BHF and APR98 with hyperons) in $\beta$-equilibrium for the same baryon densities as in Figure 4.5.

For both methods of calculation $\Sigma^-$ pops up at densities of $\sim 2 - 3$ $n_0$. Since the APR98 equation of state gives a stiffer equation of state for nucleonic matter than the corresponding BHF calculation, $\Sigma^-$ pops up at $n = 0.27$ fm$^{-3}$ for the APR98 equation of state, and at $n = 0.35$ fm$^{-3}$ for the BHF equation of state. These results agree reasonably well with calculations from the mean field theory, see amongst others, [**HH00, Gl00, PBP97**].

The introduction of hyperons gives us a significantly softer equation of state. In addition, leptons often disappear when hyperons appear. In APR98 they disappear completely, whereas in the BHF calculations it is only the muons that disappear. This is because $\Lambda$ does not appear in the BHF equation of state at the densities that are included here. For the APR98 equation of state, $\Lambda$ pops up at a density of $n = 0.67$ fm$^{-3}$. When we remember that $\mu_\Lambda = \mu_n = \mu_p + \mu_e$, and that the APR98 equation of state is stiffer due to the inclusion of three-body forces, this increases the possibilities of creating a $\Lambda$-hyperon with the APR98 equation of state. In addition to the softer equation of state, the fact that $\Lambda$ does not appear in the BHF calculations can also be due to the delicate balance between the contributions to the self energy from the baryons and from the nucleon and hyperon hole states. See diagram (a) and (b) of Figure 4.4. These contributions also contribute to shaping the properties of the baryon-baryon-interaction. Contributions to the chemical potential of the $\Lambda$ from the hole states of $\Sigma^-$, the proton and the neutron are therefore not attractive enough to lower the chemical potential of the $\Lambda$ down to the level of the neutron. Furthermore, it does not increase the chemical potential of the neutrons sufficiently, since the contributions to the self energy of the hole states of $\Sigma^-$ are attractive. We illustrate the difference between the two choices for the equation of state in Figure 4.7, through the chemical potential for different baryons for matter in $\beta$-equilibrium. We also notice that when we use the criteria of Equation (4.21), neither the $\Sigma^0$ or the $\Sigma^+$ showes up for neither the BHF or the APR98 equations of state. This is because none of the $\Sigma^0$-baryon and $\Sigma^+$-baryon interactions are sufficiently attractive. A similar argument is applied to the $\Xi^0$ and the $\Xi^-$. In the latter case the mass of the particle $\sim$ 1315 MeV, and one would need an attraction of almost 200 MeV in order to satisfy the condition $\mu_\Lambda = \mu_{\Xi^0} = \mu_n$. From the bottom panel in Figure 4.7 we nevertheless see that $\Sigma^0$ can appear at

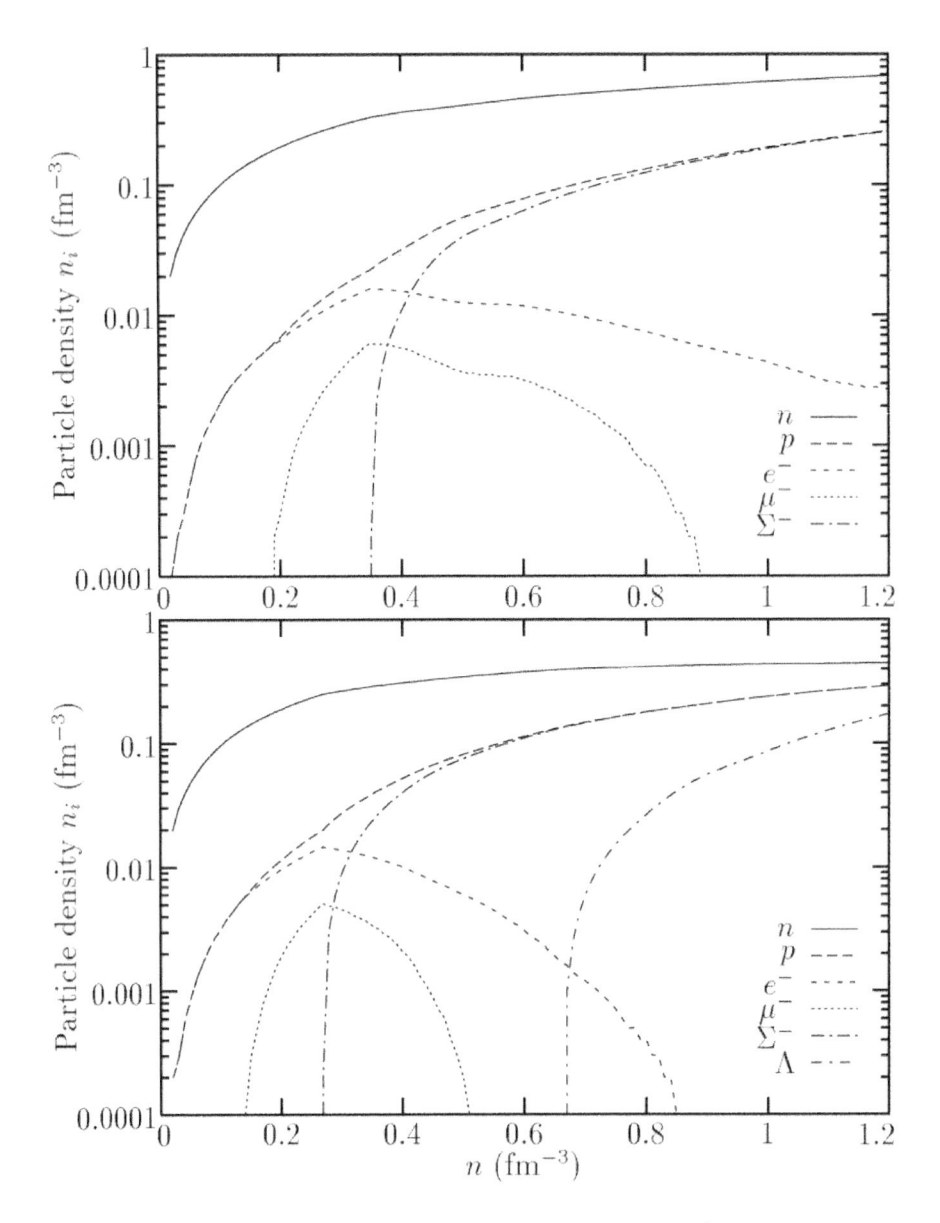

Figure 4.5: Particle densities in neutron star matter in $\beta$-equilibrium as functions of the total baryon density $n$. The upper panel represents results from the Brueckner-Hartree-Fock level with the potential from Stokes and Rijk [SR99]. In the bottom panel the nucleon part of the self-energy has been replaced with the equation of state from Equation (4.7) with $\delta = 0.2$.

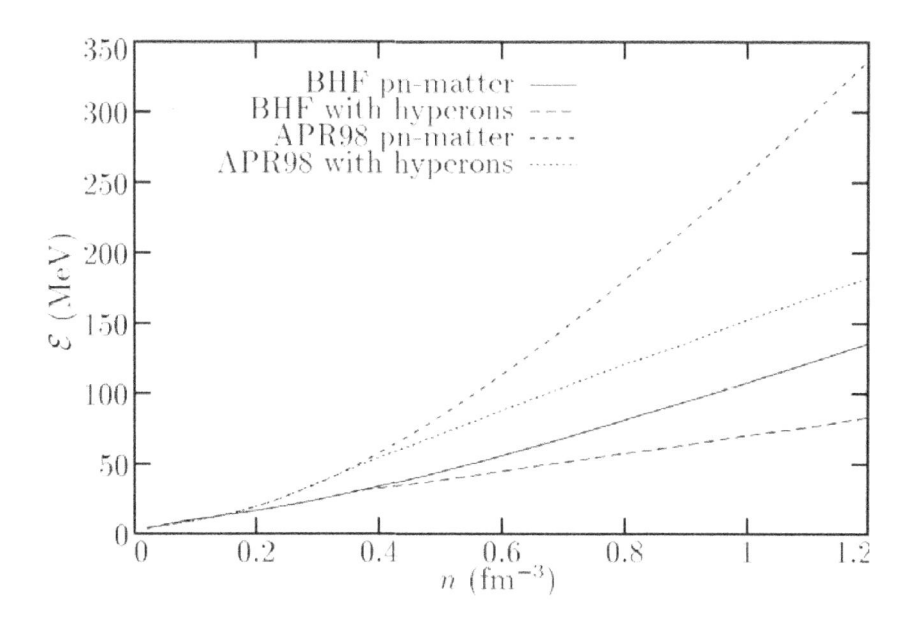

Figure 4.6: Energy per baryon in neutron star matter in $\beta$-equilibrium for different models as a function of the total baryon density $n$. See text for more details.

densities close to 1.2 fm$^{-3}$.

By using the parametrized equation of state of Akmal et al [**APR98**] from Equation (4.7) for the nucleon sector and including hyperons through the model for the baryon-baryon-interaction of the Nijmegen group [**SR99**], we thus find through many body calculations for matter in $\beta$-equilibrium that $\Sigma^-$ pops up at a density of $n = 0.27$ fm$^{-3}$, while $\Lambda$ appears at $n = 0.67$ fm$^{-3}$. Through the formation of hyperons the matter is de-leptonized at a density of $n = 0.85$ fm$^{-3}$. Within our many body approach no hyperons appear at densities below $n = 1.2$ fm$^{-3}$. Although the equation of state of Akmal et al [**APR98**] can be viewed as the currently most realistic approach to a nucleon equation of state, we must take into account the uncertainty we have in the hyperon-hyperon and hyperon-nucleon-interactions. If, for example, hyperon-hyperon interactions turn out to be more attractive than expected, this may lead to the formation of hyperons like $\Lambda$, $\Sigma^0$, $\Sigma^+$, $\Xi^-$ and $\Xi^0$ at lower densities. The hyperon-hyperon interaction, and stiffness in nucleon contribution plays a crucial role in the formation of different hyperons. These results differ from the present mean-field calculations, see for example [**Gl00, PBP97**], where all kinds of hyperons can occur at the densities that are considered here.

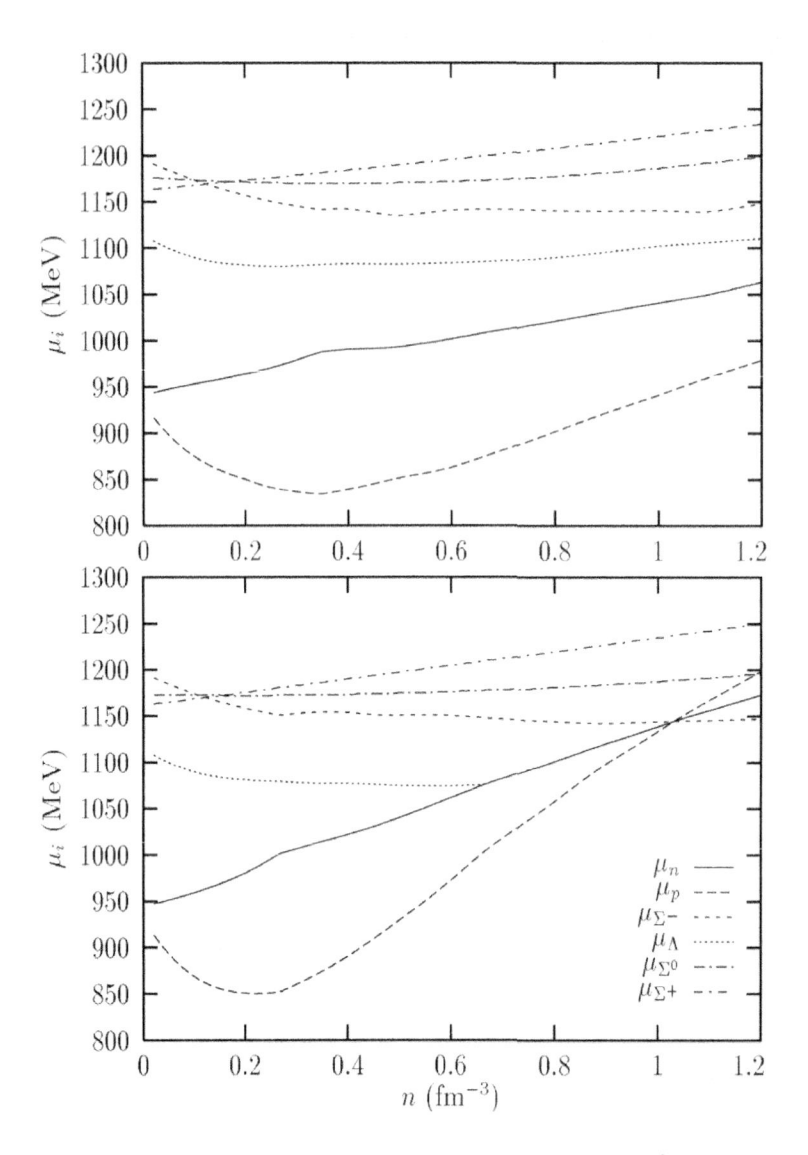

Figure 4.7: Chemical potentials in neutron star matter in $\beta$-equilibrium as a function of the total baryon density $n$. The top panel represents the results obtained at the Brueckner-Hartree-Fock level with the potential from Stokes and Rijk [SR99]. The bottom panel includes the results from the equation of state in Equation (4.7) with $\delta = 0.2$.

# Chapter 5

# Cooling processes

## 5.1 Cooling processes in neutron stars

Determination of surface temperature of neutron stars by observation of thermal black body radiation, can in principle provide us with significant information about the inner hadron matter and neutron star structure. Neutron stars are probably formed in very high internal temperatures, $T > 10^{10}$ K, in the center of a supernova explosion [**KW90**]. The dominant cooling mechanism just after formation is neutrino emissions with an initial cooling time scale of seconds. After about one day the internal temperature drops to about $10^9$ K. The neutrino processes are dominating for at least 1000 years and probably much longer (approximately 100,000 years). Photon emission takes over for neutrino cooling only when the internal temperature of the star falls to about $10^8$ K. This corresponds to a surface temperature of about $10^6$ K.

Calculations of the thermal evolution of the star, thus becomes very sensitive to the selected equation of state, the neutron star mass and radius, the magnetic field strength, the possible existence of superfluidity, pion condensation, quark matter etc. Results suggest that we first get detectable photon emissions in the soft X-ray band, and young neutron stars can thus be detected as discrete X-ray sources (different from pulsars in binary systems that emit X-rays by mass accresion from the companion star).

The first cooling period after the internal temperature has fallen below $10^{10}$ K is as stated dominated by neutron emitting cooling processes. This as any emitted neutrino freely can escape from the neutron star, without further interaction with the neutron star matter. At very high temperatures $T > 10^9$ K the dominant process is probably the direct Urca processes that go as follows

$$n \to p + e^- + \bar{\nu}_e ~,\tag{5.1}$$

or

$$e^- + p \to n + \nu_e ~.\tag{5.2}$$

We have a similar structure for other Urca-processes, like

$$\Lambda \to p + e + \bar{\nu}_e ~,\tag{5.3}$$
$$\Sigma^- \to n + e + \bar{\nu}_e ~,\tag{5.4}$$
$$\Sigma^- \to \Lambda + e + \bar{\nu}_e ~,\tag{5.5}$$
$$\Sigma^- \to \Sigma^0 + e + \bar{\nu}_e ~,\tag{5.6}$$
$$\Xi^- \to \Lambda + e + \bar{\nu}_e ~,\tag{5.7}$$
$$\Xi^- \to \Sigma^0 + e + \bar{\nu}_e ~,\tag{5.8}$$
$$\Xi^0 \to \Sigma^+ + e + \bar{\nu}_e ~,\tag{5.9}$$
$$\Xi^- \to \Xi^0 + e + \bar{\nu}_e ~,\tag{5.10}$$

and we also have similar processes in which muons replace the electrons.

Alongside these processes, we have modified Urca processes, of the type

$$n + n \to n + p + e^- + \bar{\nu}_e ~,\tag{5.11}$$
$$n + p + e^- \to n + n + \nu_e ~,\tag{5.12}$$
$$n + n \to n + p + \mu^- + \bar{\nu}_\mu ~,\tag{5.13}$$
$$n + p + \mu^- \to n + n + \nu_\mu ~.\tag{5.14}$$

As we see, these can be separated from the direct Urca processes by the presence of a companion particle which provides sufficient energy for the reaction to be carried out.

Other possible reactions are nucleon pair bremsstrahlung

$$n + p \to n + p + \nu + \bar{\nu} ~,\tag{5.15}$$

neutrino pair bremsstrahlung

$$e^- + (Z, A) \to e^- + (Z, A) + \nu + \bar{\nu} ~,\tag{5.16}$$

pion reactions

$$\pi^- + n \to n + e^- + \bar{\nu}_e ~,\tag{5.17}$$
$$\pi^- + n \to n + \mu^- + \bar{\nu}_\mu ~,\tag{5.18}$$
$$n + e^- \to n + \pi^- + \nu_e ~,\tag{5.19}$$

$$n + \mu^- \rightarrow n + \pi^- + \nu_\mu \quad , \tag{5.20}$$

and $\beta$-decay with quarks

$$d \rightarrow u + e^- + \bar{\nu}_e \quad , \tag{5.21}$$

$$u + e^- \rightarrow d + \bar{\nu}_e \quad , \tag{5.22}$$

$$s \rightarrow u + e^- + \bar{\nu}_e \quad , \tag{5.23}$$

$$u + e^- \rightarrow s + \nu_e \quad . \tag{5.24}$$

We get the photon luminosity by assuming black body-photon emission from the surface at an "effective" surface temperature. If the neutrino luminosity is known, we can then calculate the timescale for cooling and extract the temperature of a neutron star as a function of the time.

The existence of pion condensates or quark matter should accelerate the cooling dramatically. The existence of superfluid layers should reduce the heat capacity and cooling time, but also suppress neutrino producing processes which will increase the cooling time. The magnetic field should reduce the photons opacity, and increase the surface temperature and the photon luminosity. Cooling curves calculated from different models can be compared with observations from the remnant stars after supernova explosions [Øs89].

In this paper, we will now look more closely at the direct Urca process, calculate neutrino emissions for the different processes and create profiles of these for different densities.

## 5.2 Direct Urca

### 5.2.1 Direct Urca processes

The direct Urca process is assumed to be central in the cooling of neutron stars. The direct Urca processes are the simplest neutrino emitting processes, and have a general form

$$B_1 \rightarrow B_2 + l + \bar{\nu}_l \quad , \tag{5.25}$$

$$B_2 + l \rightarrow B_1 + \nu_l \quad , \tag{5.26}$$

in which $B_1$ and $B_2$ are baryons, and $l$ is a lepton. Baryons can be nucleons, but also hyperons like $\Lambda$, $\Sigma$ and $\Xi$, or isobars like $\Delta^0$ and $\Delta^-$. The leptonis either an electron or a muon. Beta equilibrium demands the chemical potentials to satisfy

$$\mu_{B_1} = \mu_{B_2} + \mu_l \ . \tag{5.27}$$

We must also satisfy conservation of momentum, which we express through the triangle inequality

$$p_{F_i} + p_{F_j} \geq p_{F_k} \ , \tag{5.28}$$

where i, j and k corresponds to $B_2$, $l$ and $B_1$. The cyclic permutations of this inequality must also be satisfied, and we thus get three such equations. When these equations are not satisfied, the modified Urca processes are the dominant mechanisms for neutrino emissions [**Pr94**].

## 5.2.2 Direct Urca with nucleons

The direct Urca process with nucleons is the simplest and most powerful neutrino process. It consists of two consecutive reactions, $\beta$-decay and $\beta$-capture,

$$n \to p + e + \bar{\nu}_e, \ n + e \to p + \nu_e \ . \tag{5.29}$$

This is the most important process in the core of a neutron star. The process is leading the nucleons into $\beta$-equilibrium, where the chemical potentials thus satisfy

$$\mu_n = \mu_p + \mu_e \ . \tag{5.30}$$

If equilibrium is not reached, the one reaction, after Le Chatelieres principle, will be more intense, and change the proton and neutron fraction towards the equilibrium value. Equilibrium exists when both reactions have the same reaction rate.

The direct Urca process thus does not change the composition of matter in $\beta$-equilibrium. The most important feature of the process is the threshold, described by the step-function

$$\theta(p_F^e + p_F^p + p_F^n) \ , \tag{5.31}$$

which expresses the fulfillment of the triangle inequality (see Equation (5.28)). The step function thus starts the direct Urca process when the density of the matter is sufficiently high, ie when the momentums satisfy this triangle inequality. In neutron star matternøytronstjernematerie $p_F^n$ is larger than $p_F^p$ and $p_F^e$, and we thus get Equation (5.31). When the density $n$ is approximately equal to the standard nuclear density $n_0$, we typically get that $p_F^n$ is about 340 MeV/c while $p_F^e$ and $p_F^p$ are approximately (60-100) MeV/c. Then the triangle inequality is not fulfilled, and the direct Urca process with nucleons is prohibited. Since $p_F^p$ and $p_F^e$ grow faster with higher density than

$p_F^n$, the process can start at slightly higher densities (sometimes $n_0$). Formally, the direct Urca process thus is allowed if the proton fraction among the baryons exceeds a certain critical value, and we shall now give an example of the calculation of this critical proton fraction with NPE-matter:

Chemical equilibrium thus gives us:

$$\mu_n = \mu_p + \mu_e \ . \tag{5.32}$$

Furthermore, we must require the star matter to be electrically neutral

$$n_e = n_p \ , \tag{5.33}$$

which gives

$$\frac{p_{F_e}^3}{3\pi^2 \hbar^3} = \frac{p_{F_p}^3}{3\pi^2 \hbar^3} \ , \tag{5.34}$$

which then again naturally gives $p_{F_e} = p_{F_p}$. The proton fraction is defined as

$$x_p = \frac{n_p}{n_n + n_p} \ . \tag{5.35}$$

This then must satisfy

$$x_p = \frac{n_p}{n_n + n_p} \geq \frac{p_{F_p}^3}{p_{F_p}^3 + (p_{F_p} + p_{F_e})^3} \ . \tag{5.36}$$

The critical proton fraction $x_p^c$, can then be expressed by

$$x_p \geq \frac{1}{1 + (\frac{p_{F_e}}{p_{F_p}})^3} \geq \frac{1}{9} \ , \tag{5.37}$$

since

$$p_{F_e} + p_{F_p} \geq p_{F_n}. \tag{5.38}$$

In the simplest models where compact matter is treated as an ideal gas of Fermi particles (see eg. [**ST83**]), the proton fraction is not sufficiently high at any densities. This does however not have to be the case for more realistic equations of state (see eg. [**LPP91**]). For some more realistic models for neutron star matter, the proton fraction exceeds the critical value at densities corresponding to a few times the standard nuclear density ($n_0$). The equations of state that provide the highest proton fraction, and thus the highest probability for the presence of the direct Urca process, are the equations of state with high symmetry energy. These will allow the direct Urca processes with nucleons in the cores of neutron stars more massive than 1.4-1.6 $M_\odot$.

The step function (5.31) thus switches on the direct Urca process in a step, in which the emissivity jumps up from zero to its final value immediately as the density reaches the critical threshold. This is of course an approximation. (This approximation comes from the phase space splitting we do in the calculations.) In reality, the direct Urca process also exists below this critical density, but it is very strongly suppressed. As the matter is very degenerate it will be reduced exponentially. The emissivity is suppressed approximately by the function $\exp(-\chi)$, where $\chi = -v_{F_p}(p_{F_n} - p_{F_p} - p_{F_e})/T$ [**YKG01**]. This effect is often referred to as thermal broadening of the direct URCA threshold. In order to qualitatively explain this effect it is sufficient to replace the step-function $\theta$ with an approximate function of the shape $(e^\chi + 1)^{-1}$. This thermal broadening mechanism still seems to be weak and unimportant for many applications.

### 5.2.3 Direct Urca with muons

We will now look at the direct Urca processes involving other particles than $n$, $p$ and $e$. If muons are present, we have the same process as described above with nucleons, only the muons now replace the electrons:

$$n \to p + \mu + \bar{\nu}_\mu, \quad \mu + p \to n + \nu_\mu \ . \tag{5.39}$$

This process then comes in addition to the process with electrons. The emissivity is given by the same equation as for the basic nucleon process above, since the condition of beta equilibrium requires that the chemical potentials for electrons and muons are equal, $\mu_\mu = \mu_e$. The difference between the processes appears at the threshold. The step-function $\theta(p_F^e + p_F^p + p_F^n)$ must, in the reaction with muons be replaced with the step-function $\theta(p_F^\mu + p_F^p + p_F^n)$. The muon process will thus start at a slightly higher density than the corresponding reaction with electrons. The emissivity once the process starts is the equal to the electron process. Once the muon process starts running, the total emissivity thus doubles.

### 5.2.4 Direct Urca with hyperons

If the equation of state of the neutron star core allows the presence of hyperons, we can also get direct Urca processes with hyperons. This means processes like

$$\Lambda \to p + e^- + \bar{\nu}_e \ ,$$

or equivalent with $\Sigma$ or other hyperons, and possibly also those processes with muons. The neutrino emissivity is calculated in exactly the same way as for the direct Urca process with nucleons. The change is that we get some other constants $C$, $f_1$ og $g_1$. We will revisit this in the calculation of the emissivity below.

# Chapter 6

# Neutrino emissivity

The matter inside a neutron star may, as we have seen earlier, contain many types of particles. In addition to nucleons, such as neutrons, protons and $\alpha$-particles, electrons and positrons, we also have the presence of hyperoner such as $\Lambda$ and $\Sigma$-particles with similarly even more possible processes. This makes the problem of finding the matter composition a very complicated one, something that again is reflected in the problem of finding the most accurate estimate of the rate of neutrino emission, and thus the cooling rate of neutron stars. When the problem of the matter composition is resolved, the rate of neutrino emissions from the beta processes can be found by calculating the sum of the contributions from all the different reactions.

## 6.1 Neutrino emissivity from the Direct Urca process

We will now derive the analytical expression for the neutrion emissivity from the direct Urca process. We will set $\hbar = c = 1$, and all momentums are four-momentums unless otherwise stated. We will assume that the neutrinos escape instantly, without interfering with the stellar matter. This allows us to simplify the angular integration by neglecting neutrinos in the $\delta$-function that preserves the four-momentum. We will also assume that $k_B T$ is much smaller than the chemical potentials of baryons and leptons. This gives us Equation (5.32) above. We should first look at a direct Urca process of the type $B_1 \to B_2 + l + \bar{\nu}_l$, where $B_1$ and $B_2$ are baryons, for example a neutron and a proton, $l$ is a lepton and $\bar{\nu}_l$ the corresponding antineutrino.

The expression for the emissivity from the direct Urca process is given by

$$\varepsilon_\nu = \prod_{i=1}^{4} \int \frac{d^3 p_i}{(2\pi)^3} E_\nu W n_{B_1}(1-n_l)(1-n_{B_2}) \,, \tag{6.1}$$

where we have

$$W = \frac{(2\pi)^4}{\prod_{i=1}^{4} 2E_i} \delta^{(4)}(p_{B_1} - p_l - p_{B_2} - p_{\nu_l}) M^2 \,. \tag{6.2}$$

Here $M$ is the Feynman matrix, $B_1$, $B_2$ and $l$ are the baryons and the lepton that parttake in the Urca process, and $n_i$ are the Fermi-Dirac distribution functions. The factors $(1-n_{B_2})$ and $(1-n_l)$ are blocking factors for the final states. When we insert the expression for $W$, we obtain for the emissivity

$$\begin{aligned}\varepsilon_\nu &= \frac{1}{2^{12}\pi^8} \int \frac{d^3 p_{B_1}}{2E_{B_1}} \int \frac{d^3 p_{B_2}}{2E_{B_2}} \int \frac{d^3 p_l}{2E_l} \int \frac{d^3 p_\nu}{2} n_{B_1}(1-n_l)(1-n_{B_2}) \\ &\times \delta^{(4)}(p_{B_1} - p_l - p_{B_2} - p_\nu) M^2 \,.\end{aligned} \tag{6.3}$$

The expression for the Feynman matrix is [**GS84**]

$$M = \frac{G_F}{\sqrt{2}} C u_{B_2}[\gamma_\mu f_1 + \gamma_\mu \gamma_5 g_1] u_{B_1} \bar{u}_l \gamma^\mu (1+\gamma_5) u_\nu \,, \tag{6.4}$$

where $C$ is an expression with the Cabibbo angle, who will be either $\cos\theta_C$ or $\sin\theta_C$ depending on which reaction we have. This depends on what quark transitions we have in the particular reaction, specifically if we have a change in strangeness $S$. $\Delta S = 0$ gives $\cos\theta_C$, while $\Delta S = 1$ gives $\sin\theta_C$. This gives us the matrix element

$$\begin{aligned}|M|^2 &= M \times M^* = [\frac{G_F}{\sqrt{2}} C [u_{B_2} \Gamma_1 u_{B_1}][\bar{u}_l \gamma^\mu \Gamma_3 u_\nu] & (6.5) \\ &\times \frac{G_F}{\sqrt{2}} C [u_{B_2} \bar{\Gamma}_2 u_{B_1}][\bar{u}_l \bar{\Gamma}_4 u_\nu] \,. & (6.6)\end{aligned}$$

By multiplying out, calculating the traces to each paragraph, and simplifying, we get finally the expression

$$\begin{aligned}M^2 &= 64 \frac{G_F^2}{2} C^2 [(p_{B_1} p_\nu)(p_{B_2} p_l)(f_1+g_1)^2 + (p_{B_1} p_l)(p_{B_2} p_\nu)(f_1-g_1)^2 \\ &\quad - m_{B_1} m_{B_2} c^2 (p_\nu p_l)(f_1^2 - g_1^2)] \,.\end{aligned} \tag{6.7}$$

We will now first briefly outline the procedure for the calculation of the approximate expression of Prakash et al [**PPL92, LPP91**], before we in the next section make a more thorough review of a more accurate calculation of the neutrino emissivity. In the approximation we are doing now, we assume that the momentum of baryons and leptons is restricted to around the Fermi surface and thus we can approximate the integral from 0 to $k_F$ with values around $k_F$.

If we further set $f_1 \approx g_1$ in the expression for the squared matrix element, the last two terms disappear, and we can rewrite the factor $(f_1 + g_1)^2$ in the first term as $(f_1^2 + 3g_1^2)$, so that we are left with the expression

$$M^2 = 64 \frac{G_F^2}{2} C^2 (p_{B_1} p_\nu)(p_{B_2} p_l) \left( f_1^2 + 3g_1^2 \right) \quad . \tag{6.8}$$

This corresponds to Equation (3.5) in [**Iw82**]. We then insert this in to Equation (6.3) to calculate the emission rate. We first integrate over the angles. In this approximation we assume angle independence and get for the angular integral

$$\frac{(4\pi)^3}{2 \left| \mathbf{p}_{B_2} \right| \left| \mathbf{p}_{B_1} \right| \left| \mathbf{p}_l \right|} \quad . \tag{6.9}$$

See also the next section for a more detailed calculation of this. When we later use the relationships $E_\nu = p_\nu$ for neutrinos, and $E_j dE_j = p_j dp_j$ for the other particles, this cancels the denominator.

The squared matrix element summed over the spins of the end particles and averaged over the angles after our approximations thus gives us a factor $32 G_F^2 C^2 \left( f_1^2 + 3g_1^2 \right) (p_{B_1} p_\nu)(p_{B_2} p_l)$. Moreover, we can now rewrite

$$\begin{aligned} p_{B_1} p_\nu &= E_{B_1} E_\nu - \mathbf{p}_{B_1} \mathbf{p}_\nu \ , \\ p_{B_2} p_l &= E_{B_2} E_l - \mathbf{p}_{B_2} \mathbf{p}_l \ , \end{aligned} \tag{6.10}$$

where we non-relativisticly can approximate by deleting the last terms. We are then left with $E_{B_1} E_\nu E_{B_2} E_l$. $E_{B_1}$ then gives us then approximativly $m_{B_1}$ and $E_{B_2}$ gives us $m_{B_2}$.

For the integral over $E_l E_\nu^3 n_{B_1} (1 - n_l)(1 - n_{B_2})$ we will now follow the procedure in Appendix F of [**ST83**] rather slavishly. We insert for the Fermi-Dirac functions, and the integral over these can be calculated by making a substitution of variables as follows:

$$\begin{aligned} x_{B_1} &= \beta(E_{B_1} - \mu_{B_1}) \ , \\ x_{B_2} &= -\beta(E_{B_2} - \mu_{B_2}) \ , \\ x_l &= -\beta(E_l - \mu_l) \ . \end{aligned} \tag{6.11}$$

With $\beta \equiv 1/(k_B T)$, and where $\mu_{B_1}$, $\mu_{B_2}$ and $\mu_l$ are the chemical potentials of the different particles. At the same time we then define

$$y = \frac{E_\nu}{k_B T} \quad . \tag{6.12}$$

We then get the $1/\beta^6$, which gives us a factor $(k_B T)^6$. $E_l$ gives us a factor $(x_l + \mu_l)$, where $\mu_l$ will be dominating, and we then set this term approximatively equal to $\mu_l$. We see that we now have an expression

$$\frac{1}{2^{12}\pi^8} 64 \frac{G_F^2}{2} C^2 \left(f_1^2 + 3g_1^2\right) \frac{(4\pi)^3}{2} m_{B_1} m_{B_2} \mu_l (k_B T)^6 \tag{6.13}$$

which we can put outside the integration limits, and it then only remains to integrate over $E_\nu^3 n_{B_1}(1 - n_l)(1 - n_{B_2})$, which after substitution gives us the expression

$$y^3 \left(1 - e^{x_{B_1}}\right)^{-1} \left(1 - e^{x_{B_2}}\right)^{-1} \left(1 - e^{x_l}\right)^{-1} \times \delta(x_{B_1} - x_l - x_{B_2} - y) \quad , \tag{6.14}$$

To solve this we use residue-calculations. We split the integral as follows:

$$I \equiv \int_0^\infty dy\, y^3\, J \quad , \tag{6.15}$$

where

$$J \equiv \int \prod_{j=B_1,B_2,l} dx_j \left(1 - e^j\right)^{-1} \delta\left[\sum_{j=B_1,B_2,l} x_j - y\right] \quad . \tag{6.16}$$

The calculation will be exactly as in Appendix F in [ST83]. The only difference is that we in this calculation must calculate

$$\text{Residue ved } z = 0 \text{ av } \left[ e^{-izy} \left(\frac{\pi}{\sinh \pi z}\right)^3 \right] \quad , \tag{6.17}$$

while Shapiro and Teukolsky in [**ST83**] perform calculations for the modified Urca process in which two more particles are involved, and thus instead get a factor $(\pi/\sinh \pi z)^5$. By calculating (6.17) you get the expression

$$\frac{\pi^2}{2} + \frac{y^2}{2} \quad , \tag{6.18}$$

which gives us the following equation to find $I$:

$$I = \int_0^\infty dy \left(\frac{\pi^2}{2} y^3 + \frac{1}{2} y^5\right) (e^y + 1)^{-1} \quad . \tag{6.19}$$

We solve this by using Maple (see Appendix B), and get

$$I = \frac{457}{4040} \pi^6 \quad . \tag{6.20}$$

The reverse process, electron capture, will give the same luminosity, but neutrinos instead of antineutrinos. The total the emissivity will then be twice as big, and give us a factor 2, see discussion in Chapter 5.2.2. The sum of all this gives us

$$\varepsilon_\nu = \frac{457\pi}{10080} G_F^2 C^2 \left(f_1^2 + 3g_1^2\right) m_{B_1} m_{B_2} \mu_l (k_B T)^6 \ . \tag{6.21}$$

The Neutrino emissivity is given in units of energy per volume per time, MeV fm$^{-3}$ s$^{-1}$. When we do a dimensional analysis, we see that the factor $G_F^2$ contributes with MeV$^2$ (m$^3$)$^2$, while $m_{B_1} m_{B_2}$ each contribute with with MeV $c^{-2}$, ie. MeV s$^2$ m$^{-2}$. The chemical potential $\mu_l$ contributes with MeV, and the factor $(k_B T)^6$ provides us with MeV$^6$. We then get the total ofMeV$^{11}$m$^2$ s$^4$. When we the have $\hbar$ in the units MeV s, and $c$ in m s$^{-1}$, we see that we need to insert $\hbar^{-10} c^{-5}$ into the expression to get the correct unit, and we end up with the equation

$$\varepsilon_\nu = \frac{457\pi}{10080} G_F^2 C^2 \left(f_1^2 + 3g_1^2\right) \frac{m_{B_1} m_{B_2} \mu_l}{\hbar^{10} c^5} (k_B T)^6 \theta(p_F^l + p_F^{B_2} - p_F^{B_1}) \ , \tag{6.22}$$

in which the step function $\theta(p_F^e + p_F^p - p_F^n)$ tests the triangle inequality, whether the Fermi levels of the particles in the process are such that the current process can go. In Table 6.1 we find the value of the various constants used in Equation (6.22) for the different reactions.

### Table 6.1
The values of the constants $C$, $f_1$ and $g_1$ in the various weak processes for nucleons and hyperons.

| $Transition$ | $C$ | $f_1$ | $g_1$ |
|---|---|---|---|
| $n \to p l \bar{\nu}_l$ | $\cos\theta_C$ | $1$ | $F + D$ |
| $\Lambda \to p l \bar{\nu}_l$ | $\sin\theta_C$ | $-\sqrt{3/2}$ | $-\sqrt{3/2}(F + D/3)$ |
| $\Sigma^- \to n l \bar{\nu}_l$ | $\sin\theta_C$ | $-1$ | $-(F - D)$ |
| $\Sigma^- \to \Lambda l \bar{\nu}_l$ | $\cos\theta_C$ | $0$ | $\sqrt{3/2} D$ |
| $\Sigma^- \to \Sigma^0 l \bar{\nu}_l$ | $\cos\theta_C$ | $\sqrt{2}$ | $\sqrt{2} F$ |
| $\Xi^- \to \Lambda l \bar{\nu}_l$ | $\sin\theta_C$ | $\sqrt{3/2}$ | $\sqrt{3/2}(F - D/3)$ |
| $\Xi^- \to \Sigma^0 l \bar{\nu}_l$ | $\sin\theta_C$ | $\sqrt{1/2}$ | $(F + D)\sqrt{2}$ |
| $\Xi^0 \to \Sigma^+ l \bar{\nu}_l$ | $\sin\theta_C$ | $1$ | $F + D$ |
| $\Xi^- \to \Xi^0 l \bar{\nu}_l$ | $\cos\theta_C$ | $1$ | $F - D$ |

Here we use the values $\sin\theta_C = 0.231 \pm 0.003$, $F = 0.477 \pm 0.012$, $D = 0.756 \pm 0.011$.

If we then rewrite the equation (6.22) specifically for the direct Urca process with nucleons, we obtain $C^2 = \cos^2\theta_c$ and $f_1^2 = 1$, which gives us the expression

$$\varepsilon_\nu = \frac{457\pi}{10080} G_F^2 \cos^2\theta_c \left(1 + 3g_1^2\right) \frac{m_n m_p \mu_e}{\hbar^{10} c^5} (k_B T)^6 \theta(p_F^e + p_F^p - p_F^n) \ . \quad (6.23)$$

Equation (6.23) now corresponds to the term we have for direct Urca with nucleons from Prakash et al. [**PPL92, LPP91**]. This approximation assumes that the particles included are free and the approximation can be improved by eg. replacing the bare masses in Equation (6.22) with effective masses.

## 6.2 A more exact calculation of the neutrino emissivity

We will now undertake a more thorough review of the calculations of the neutrino emissivity where we keep the last terms, and perform integrations also over these. During the angular integrations that follow, we follow the procedure in Appendix F in [**ST83**]. We look first at the integral over the angles $\Omega$. These are here given by

$$\int \Omega_{B_1} \int d\Omega_\nu \int d\Omega_l \int d\Omega_{B_2} \delta^{(3)} \left(\mathbf{p}_{B_1} - \mathbf{p}_l - \mathbf{p}_{B_2} - \mathbf{p}_\nu\right) \ . \quad (6.24)$$

We re-write the $\delta$-function as

$$\frac{\delta\left(|\mathbf{p}_{B_2}| - |\mathbf{p}_{B_1} - \mathbf{p}_l - \mathbf{p}_\nu|\right)}{|\mathbf{p}_p|^2} \delta\left(\Omega_{B_2} - \Omega_{B_1 - l - \nu}\right) \ . \quad (6.25)$$

The integral over $\Omega_{B_2}$ will then give us 1, and we get

$$\int d\Omega_{B_1} \int d\Omega_\nu \int d\Omega_l \frac{\delta\left(|\mathbf{p}_{B_2}| - |\mathbf{p}_{B_1} - \mathbf{p}_l - \mathbf{p}_\nu|\right)}{|\mathbf{p}_p|^2} \ . \quad (6.26)$$

We the define the variable

$$x = \mathbf{p}_{B_1} - \mathbf{p}_\nu \ . \quad (6.27)$$

Here we have chosen the z-axis for $\mathbf{p}_l$ along $x$. By using the identity $\delta[f(x)] = \delta(x-a)/|f'(a)|$, where $f(a) = 0$, and inserting $|\mathbf{p}_i| = p_i$, we get for the last $\delta$-function

$$\delta(|\mathbf{p}_{B_2}| - |\mathbf{p}_{B_1} - \mathbf{p}_l - \mathbf{p}_\nu|) = \frac{\delta(\cos\theta_{xl} - (p_{B_2}^2 - x^2 - p_l^2)/(2xp_l))}{xp_l/p_{B_2}} \ . \quad (6.28)$$

This gives us for the angular integral

$$\frac{2\pi}{|\mathbf{p}_{B_2}||\mathbf{p}_l|} \int d\Omega_{B_1} \int d\Omega_\nu \frac{1}{|\mathbf{p}_{B_1} - \mathbf{p}_\nu|} \quad . \tag{6.29}$$

We will now assume that $|\mathbf{p}_{B_1}| \gg |\mathbf{p}_\nu|$. This approximation will generally hold, but for example in a supernova environment, where neutrinos can become very energetic, it will no longer be valid. The integral above, under this assumption can be rewritten to

$$\frac{2\pi}{|\mathbf{p}_{B_2}||\mathbf{p}_{B_1}||\mathbf{p}_l|} \int d\Omega_{B_1} \int d\Omega_\nu \quad . \tag{6.30}$$

The angulare integration will accordingly depend on the angular dependence of the squared Feynman amplitude. If there is no angular dependence, as predicted by Shapiro and Teukolsky [**ST83**], and most others, the integral will be trivial, and just give us

$$\frac{(4\pi)^3}{2|\mathbf{p}_{B_2}||\mathbf{p}_{B_1}||\mathbf{p}_l|} \quad . \tag{6.31}$$

If we however are to assume an angular dependence in the expression of the Feynman matrix in (6.7), we must perform the integration explicitly. We consider the first term in the equation, and for the moment omit the constants in front. We rewrite the first of the terms using the condition of conservation of momentum ($p_{B_1} - p_l - p_{B_2} - p_\nu = 0$), and we get

$$p_{B_1}^2 + p_\nu^2 - 2p_{B_1}p_\nu = p_l^2 + p_{B_2}^2 + 2p_l p_{B_2} \quad . \tag{6.32}$$

By using the bare masses we then get

$$p_l p_{B_2} = -p_{B_1} p_\nu + \frac{m_{B_1}^2 - m_{B_2}^2 - m_l^2}{2} \quad . \tag{6.33}$$

When we then insert this for $(p_{B_1}p_\nu)(p_{B_2}p_l)$ in the expression for the squared Feynman Matrix in (6.7) we get

$$(p_{B_1}p_\nu)(p_{B_2}p_l) = (p_{B_1}p_\nu)(-p_{B_1}p_\nu + \frac{m_{B_1}^2 - m_{B_2}^2 - m_l^2}{2}) \quad . \tag{6.34}$$

We then use the relationship

$$p_{B_1}p_\nu = E_{B_1}E_\nu - \mathbf{p}_{B_1}\mathbf{p}_\nu = E_{B_1}E_\nu - p_{B_1}p_\nu \cos\theta_{B_1\nu} \quad , \tag{6.35}$$

and see that the expression for $(p_{B_1}p_\nu)(p_{B_2}p_l)$ in Equation (6.34) is proportional with

$$c_0 + c_1 \cos\theta_{n\nu} + c_2(\cos\theta_{n\nu})^2 \ . \tag{6.36}$$

When we then do the integration over $\Omega_\nu$ in Equation (6.30) we see that the term with $c_1$ becomes zero, and we are left with

$$\frac{2\pi}{|\mathbf{p}_{B_2}||\mathbf{p}_{B_1}||\mathbf{p}_l|}\int d\Omega_{B_1}\int d\Omega_\nu (p_{B_1}p_\nu)(p_{B_2}p_l) = \frac{(4\pi)^3}{2|\mathbf{p}_{B_2}||\mathbf{p}_{B_1}||\mathbf{p}_l|}(c_0 + \frac{1}{3}c_2), \tag{6.37}$$

where we have

$$c_0 = \frac{m_{B_1}^2 - m_{B_2}^2 - m_l^2}{2} E_{B_1}E_\nu - (E_{B_1}E_\nu)^2 \ , \tag{6.38}$$

and

$$c_2 = -(p_{B_1}p_\nu)^2 = -(E_n^2 - m_n^2)E_\nu^2 \ . \tag{6.39}$$

Here the neutrino mass is set to zero, and

$$E_i = \sqrt{p_i^2 + m_i^2} \ . \tag{6.40}$$

The angular integral then finally gives us

$$\frac{(4\pi)^3}{2|\mathbf{p}_{B_2}||\mathbf{p}_{B_1}||\mathbf{p}_l|}$$
$$\times(\frac{m_{B_1}^2 - m_{B_2}^2 - m_l^2}{2} E_{B_1}E_\nu - (E_{B_1}E_\nu)^2 - \frac{1}{3}(E_n^2 - m_n^2)E_\nu^2) \ . \tag{6.41}$$

By including the constants in front of the squared Feynman matrix, we obtain for the first term:

$$(f_1 + g_1)^2 64C^2 \frac{G_F^2}{2} \frac{(4\pi)^3}{2|\mathbf{p}_{B_2}||\mathbf{p}_{B_1}||\mathbf{p}_l|}$$
$$\times(\frac{m_{B_1}^2 - m_{B_2}^2 - m_l^2}{2} E_{B_1}E_\nu - (E_{B_1}E_\nu)^2 - \frac{1}{3}(E_n^2 - m_n^2)E_\nu^2) \ . \tag{6.42}$$

Similarly, for the second term:

$$(f_1 - g_1)^2 64C^2 \frac{G_F^2}{2} \frac{(4\pi)^3}{2|\mathbf{p}_{B_2}||\mathbf{p}_{B_1}||\mathbf{p}_l|}$$
$$\times(\frac{m_{B_1}^2 - m_{B_2}^2 - m_l^2}{2} E_{B_1}E_\nu - (E_{B_1}E_\nu)^2 - \frac{1}{3}(E_n^2 - m_n^2)E_\nu^2) \ . \tag{6.43}$$

In the final term the integration over the angles gives us zero, and we get a simpler expression:

$$(f_1^2 - g_1^2)64C^2 \frac{G_F^2}{2} \frac{(4\pi)^3}{2|\mathbf{p}_{B_2}||\mathbf{p}_{B_1}||\mathbf{p}_l|} m_p m_n E_l E_\nu \ . \qquad (6.44)$$

We now move on to the integration over the momentums. First we take a look at the first term. The integral to be performed is

$$\begin{aligned}\varepsilon_{\nu_1} &= \frac{1}{2^{12}\pi^8} \int \frac{d^3 p_{B_1}}{2E_{B_1}} \int \frac{d^3 p_{B_2}}{2E_{B_2}} \int \frac{d^3 p_l}{2E_l} \int \frac{d^3 p_\nu}{2} n_{B_1}(1-n_l)(1-n_{B_2}) \\ &\times \delta^{(4)}(E_{B_1} - E_l - E_{B_2} - E_\nu) \times (f_1+g_1)^2 64 C^2 \frac{G_F^2}{2} \frac{(4\pi)^3}{2|\mathbf{p}_{B_2}||\mathbf{p}_{B_1}||\mathbf{p}_l|} \\ &\times (\frac{m_{B_1}^2 - m_{B_2}^2 - m_l^2}{2} E_{B_1} E_\nu - (E_{B_1} E_\nu)^2 - \frac{1}{3}(E_{B_1}^2 - m_{B_1}^2) E_\nu^2). (6.45)\end{aligned}$$

By using $E_\nu = p_\nu$, and for the other particles $E_j dE_j = p_j dp_j$, we cansel out the numerator $|\mathbf{p}_{B_2}||\mathbf{p}_{B_1}||\mathbf{p}_l|$, and get

$$\begin{aligned}\varepsilon_{\nu_1} &= \frac{1}{2^{11}\pi^5} \int dE_{B_1} \int dE_{B_2} \int dE_l \int E_\nu^2 dE_\nu \, n_{B_1}(1-n_l)(1-n_{B_2}) \\ &\times \delta^{(4)}(E_{B_1} - E_l - E_{B_2} - E_\nu) \times (f_1+g_1)^2 64 C^2 \frac{G_F^2}{2} \\ &\times (\frac{m_{B_1}^2 - m_{B_2}^2 - m_l^2}{2} E_{B_1} E_\nu - (E_{B_1} E_\nu)^2 - \frac{1}{3}(E_{B_1}^2 - m_{B_1}^2) E_\nu^2) \ .\end{aligned}$$
(6.46)

Når vi så setter inn for Fermi-Dirac-funksjonene, får vi:

$$\begin{aligned}\varepsilon_{\nu_1} &= (f_1+g_1)^2 64 C^2 \frac{G_F^2}{2} \frac{1}{2^{11}\pi 5} \int dE_{B_1} \int dE_{B_2} \int dE_l \int E_\nu^2 dE_\nu \\ &\times \delta(E_{B_1} - E_l - E_{B_2} - E_\nu) \\ &\times (\frac{m_{B_1}^2 - m_{B_2}^2 - m_l^2}{2} E_{B_1} E_\nu - (E_{B_1} E_\nu)^2 - \frac{1}{3}(E_{B_1}^2 - m_{B_1}^2) E_\nu^2) \\ &\times \frac{1}{1+\exp\beta(E_{B_1} - \mu_{B_1})} \times \frac{1}{1+\exp-\beta(E_l - \mu_l)} \\ &\times \frac{1}{1+\exp-\beta(E_{B_2} - \mu_{B_2})} \ .\end{aligned}$$
(6.47)

We now introduce the new variables $\frac{x_i}{\beta} = E_i - \mu_i$ and $x_\nu = E_\nu \beta$ and define

$$A_1(x_{B_1}, x_v) = (\frac{m_{B_1}^2 - m_{B_2}^2 - m_l^2}{2}E_{B_1}E_v - (E_{B_1}E_v)^2 - \frac{1}{3}(E_{B_1}^2 - m_{B_1}^2)E_v^2). \tag{6.48}$$

This gives us

$$\begin{aligned}\varepsilon_{\nu_1} &= (f_1+g_1)^2 C^2 \frac{G_F^2}{2^6 \pi^5} \int dx_{B_1} \int dx_{B_2} \int dx_l \int x_\nu^2 dx_\nu \left(\frac{1}{\beta}\right)^6 \\ &\quad \times \delta(x_{B_1} - x_l - x_{B_2} - x_\nu) A_1(x_{B_1}, x_\nu) \\ &\quad \times \frac{1}{1+e^{x_{B_1}}} \frac{1}{1+e^{-x_{B_2}}} \frac{1}{1+e^{-x_l}}.\end{aligned} \tag{6.49}$$

We now use the $\delta$-function to set $x_l = x_{B_1} - x_{B_2} - x_\nu$. When we then integrate over $x_l$ we get

$$\begin{aligned}\varepsilon_{\nu_1} &= (f_1+g_1)^2 C^2 \frac{G_F^2 (k_B T)^6}{2^6 \pi^5} \int dx_{B_1} \int dx_{B_2} \int x_\nu^2 dx_\nu \\ &\quad \times A_1(x_{B_1}, x_\nu) \\ &\quad \times \frac{1}{1+e^{x_{B_1}}} \frac{1}{1+e^{-x_{B_2}}} \frac{1}{1+e^{x_{B_2}+x_\nu - x_{B_1}}},\end{aligned} \tag{6.50}$$

where $\left(\frac{1}{\beta}\right)^6$ gives us $(k_B T)^6$. This expression is the same for the second term of Equation (6.43), apart from that we in Equation (6.43) integrate over $x_n$, and that we now get a function $A_2$ depending on $x_{B_2}$ and $x_\nu$. This gives us

$$\begin{aligned}\varepsilon_{\nu_2} &= (f_1-g_1)^2 C^2 \frac{G_F^2 (k_B T)^6}{2^6 \pi^5} \int dx_{B_1} \int dx_{B_2} \int x_\nu^2 dx_\nu \\ &\quad \times A_2(x_{B_2}, x_\nu) \\ &\quad \times \frac{1}{1+e^{x_{B_1}}} \frac{1}{1+e^{-x_{B_2}}} \frac{1}{1+e^{x_{B_2}+x_\nu - x_{B_1}}},\end{aligned} \tag{6.51}$$

where

$$A_2(x_{B_2}, x_\nu) = (\frac{m_{B_1}^2 - m_{B_2}^2 + m_l^2}{2}E_{B_2}E_v - (E_{B_2}E_v)^2 - \frac{1}{3}(E_{B_2}^2 - m_{B_1}^2)E_v^2). \tag{6.52}$$

The third and last part is easier, since our function $A_3$ here only depends on

$x_l$. For the third term the integral to be solved is

$$\begin{aligned}\varepsilon_{\nu_3} = & -(f_1^2 - g_1^2)64C^2\frac{G_F^2}{2}\frac{1}{2^{11}\pi^5}\int dE_{B_1}\int dE_{B_2}\int dE_l\int E_\nu^2 dE_\nu \\ & \times \delta(E_{B_1} - E_l - E_{B_2} - E_\nu) \times (m_{B_1} m_{B_2} E_l E_\nu) \\ & \times \frac{1}{1+\exp\beta(E_{B_1}-\mu_{B_1})} \times \frac{1}{1+\exp-\beta(E_l-\mu_l)} \\ & \times \frac{1}{1+\exp-\beta(E_{B_2}-\mu_{B_2})}, \end{aligned}$$ (6.53)

which in a similar manner gives

$$\begin{aligned}\varepsilon_{\nu_3} = & -(f_1^2-g_1^2)C^2\frac{G_F^2\,(k_BT)^7\,m_{B_1}m_{B_2}}{2^6\pi^5}\int dx_{B_1}\int dx_l\int x_\nu^3 dx_\nu \\ & \times A_3(x_l) \times \frac{1}{1+e^{x_{B_1}}}\frac{1}{1+e^{x_l+x_\nu-x_{B_1}}}\frac{1}{1+e^{-x_l}},\end{aligned}$$ (6.54)

where we have integrated over $x_{B_2}$ instead of $x_l$ as in the previous two integrations and where $A_3(x_l) = [(E_l - \mu)\beta]$.

Then, to make the next integration, we seek to integrate over $x_{B_2}$, and with $u = x_\nu - x_{B_1}$ we get for $\varepsilon$ an integral of the type

$$\int \frac{1}{1+e^{-x_{B_2}}}\frac{1}{1+e^{x_{B_2}+u}},$$ (6.55)

which we solve using Maple (se Appendix B), and find

$$\int \frac{1}{1+e^{-x_{B_2}}}\frac{1}{1+e^{x_{B_2}+u}} = \frac{\ln\left(1+e^{x_{B_2}}e^u\right) - \ln\left(e^{x_{B_2}}+1\right)}{e^u - 1}.$$ (6.56)

When we set the integration limits from 0 to the Fermi level $E_F^{B_2}$, we get the expression:

$$\begin{aligned}\varepsilon_{\nu_1} = & (f_1+g_1)^2C^2\frac{G_F^2\,(k_BT)^6}{2^6\pi^5}\int dx_{B_1}\int x_\nu^2 dx_\nu \\ & \times A_1(x_{B_1}, x_\nu) \times \frac{1}{1+e^{x_{B_1}}} \\ & \times \frac{\ln\left[2\left(1+e^{E_F^{B_2}}e^u\right)\right] - \ln\left[\left(e^{E_F^{B_2}}+1\right)(1+e^u)\right]}{e^u - 1}.\end{aligned}$$ (6.57)

Similarly we get an expression for $\varepsilon_{\nu_2}$ of the type

$$\int \frac{1}{1+e^{x_{B_1}}}\frac{1}{1+e^{-x_{B_1}+v}},$$ (6.58)

where $v = x_{B_2} + x_\nu$. For this expression we get

$$\int \frac{1}{1+e^{x_{B_1}}} \frac{1}{1+e^{-x_{B_1}+v}} = \frac{\ln\left(e^{x_{B_1}}+1\right) - \ln\left(e^{x_{B_1}}+e^v\right)}{e^v - 1} . \tag{6.59}$$

When we then integrate over $x_{B_1}$ and insert for the integration limits, we get for $\varepsilon_{\nu_2}$:

$$\begin{aligned}\varepsilon_{\nu_2} =& (f_1 - g_1)^2 C^2 \frac{G_F^2 (k_B T)^6}{2^6 \pi^5} \int dx_{B_2} \int x_\nu^2 dx_\nu \\ & \times A_2(x_{B_2}, x_\nu) \times \frac{1}{1+e^{-x_{B_2}}} \\ & \times \frac{\ln\left(1+e^{E_F^{B_1}}\right) + \ln\left(1+e^v\right) - \ln\left[2\left(e^{E_F^{B_1}} + e^v\right)\right]}{e^v - 1} .\end{aligned} \tag{6.60}$$

For $\varepsilon_{\nu_3}$ we get the same integral as for $\varepsilon_{\nu_2}$ when we integrate over $x_{B_1}$ and set $w = x_l + x_\nu$

$$\begin{aligned}\varepsilon_{\nu_3} =& -(f_1^2 - g_1^2) C^2 \frac{G_F^2 (k_B T)^7 m_{B_1} m_{B_2}}{2^6 \pi^5} \int dx_l \int x_\nu^3 dx_\nu \times A_3(x_l) \\ & \times \frac{1}{1+e^{-x_l}} \times \frac{\ln\left(1+e^{E_F^{B_1}}\right) + \ln\left(1+e^w\right) - \ln\left[2\left(e^{E_F^{B_1}} + e^w\right)\right]}{e^w - 1} .\end{aligned} \tag{6.61}$$

As mentioned in the introduction of this chapter, this is the emissivity of direct Urca process of the type $B_1 \to B_2 + l + \bar{\nu}_l$. In addition to this, we of course also have the corresponding processes of the type $B_2 + l \to B_1 + \nu_l$. This will of course give us different equilibrium conditions, which generates a different $\delta$-function of the type $\delta(E_{B_2} + E_l - E_{B_1} - E_\nu)$. This will give us three equations $\varepsilon_{\nu_1}^*$, $\varepsilon_{\nu_2}^*$ and $\varepsilon_{\nu_3}^*$, which are identical to those above, except that the factors $u$, $v$ and $w$ must now be replaced with $u^*$, $v^*$ and $w^*$, given by

$$u^* = -x_{B_1} - x_\nu , \tag{6.62}$$

$$v^* = x_{B_2} - x_\nu , \tag{6.63}$$

and

$$w^* = x_l - x_\nu . \tag{6.64}$$

The total emissivity from the direct Urca process, will then be given by the sum of the terms,

$$\varepsilon_\nu = \varepsilon_{\nu_1} + \varepsilon_{\nu_2} + \varepsilon_{\nu_3} + \varepsilon_{\nu_1}^* + \varepsilon_{\nu_2}^* + \varepsilon_{\nu_3}^* . \tag{6.65}$$

These calculations provide a certain overview of which equations must be the basis for a more accurate calculation of the neutrino emissivity of direct Urca processes. These equations can be used as a basis for numerical integrations to find more exact expression for the emission rates in the various processes.

# Chapter 7

# Results

We will now look at, compare and interpret the results we get for particle fractions for the ideal gas approximation, and the more accurate equation of state where we have included expressions of two- and three-body interactions as mentioned earlier. We will base our analysis on the composition of matter for the state equation from [**VPR00**], a model that includes two-body and three-body interactions between nucleons. We then look at the emissivity for the two cases, the ideal gas and the one with interactions, both calculated using Equation (6.22), and look at the relationship between the matter compositions and between the matter compositions and the particle fractions. To get better insight into the consequensest the various interactions in the equation of state have for the particle composition and the neutrino emissivity, we have also included a third approximation where we take only two-body interactions into consideration, and neglect the contribution of three-body interactions. We will also look at the chemical potential and link this to the neutrino emissivity.

Let us first look at the composition of matter in our ideal gas approximation.We see in Figure 7.1 how $\Sigma^-$ shows up at a density of 0.65 fm$^{-3}$, while $\Lambda$ does not show up until at a density above 1.1 fm$^{-3}$, and therefore does not contribute in the density range we examine here. $\Sigma^-$ is energetically favorable here because of the high chemical potential we get for the electron (see Figure 7.4). We here note that the Ferm gas model dramatically underestimates the fraction of electrons at the start of the density spectrum, while the opposite is the case when reach densities above 0.5 fm$^{-3}$. The muon will in this case not appear in the Fermi gas model until the density approaches 0.5 fm$^{-3}$. In models that include interactions between nucleons, the muon will appear already around the normal nucleon density of 0.16 fm$^{-3}$ and will thus be with us from the start of the density spectrum we see in Figure 7.3.

We then look at the matter composition in Figure 7.3, where we have

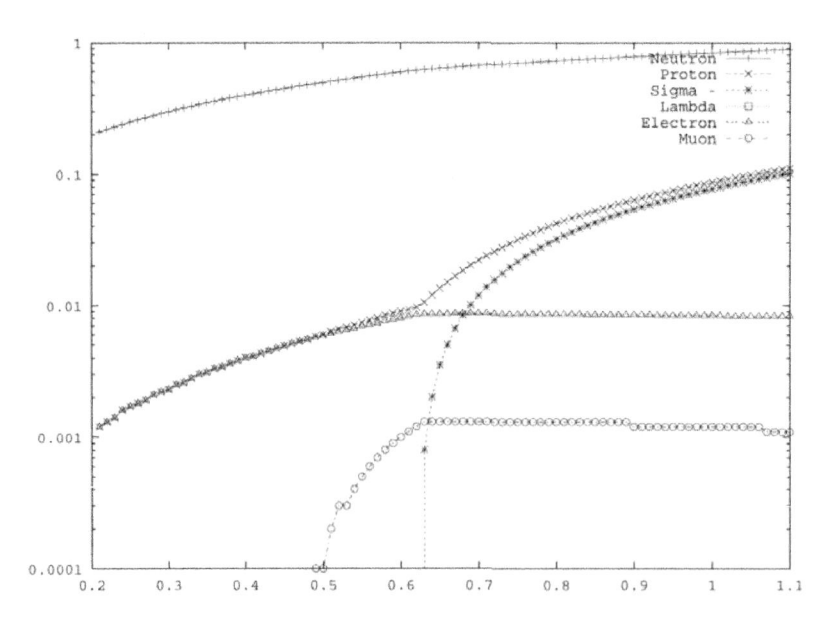

Figure 7.1: Here we see the composition of matter in the ideal gas approximation. Particle fractions of neutrons, protons, electrons, muons, $\Sigma^-$ and $\Lambda$ are plotted along a logaritmic axis as a function of the total density which is given in terms of fm$^{-3}$.

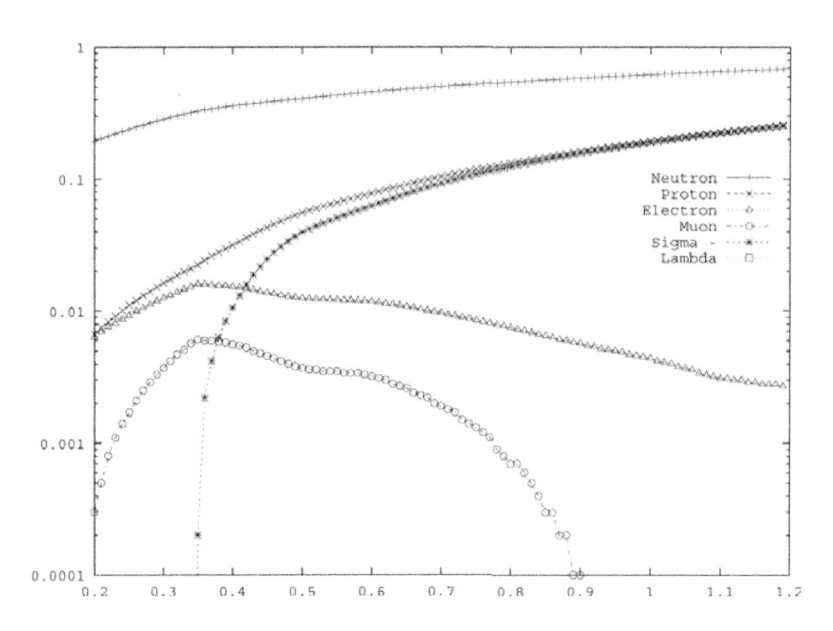

Figure 7.2: Here we see the composition of matter for the equation of state that include two-body, but not three-body interactions. Particle fractions of neutrons, protons, electrons, muons, $\Sigma^-$ and $\Lambda$ are plotted along a logaritmic axis as a function of the total density which is given in terms of fm$^{-3}$.

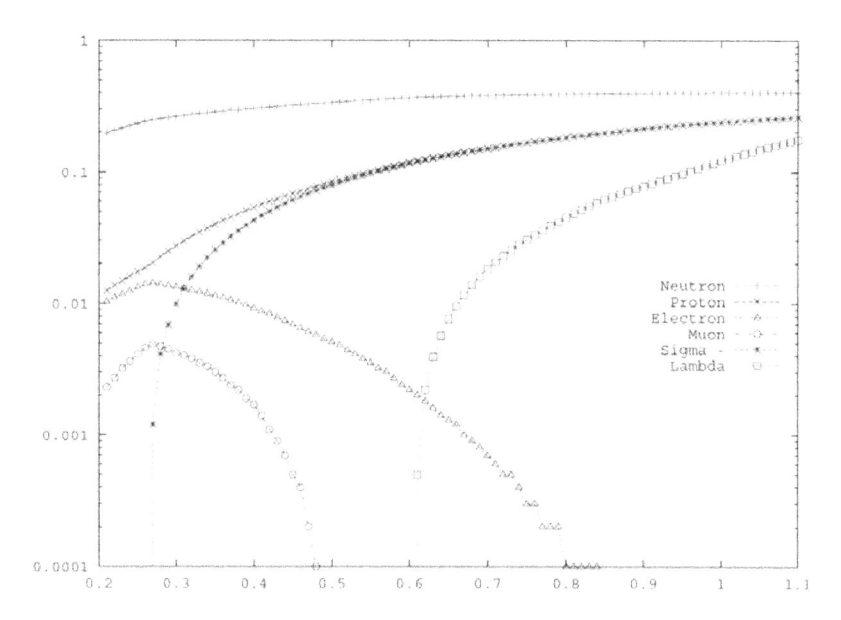

Figure 7.3: Here we see the composition of matter for the equation of state from [VPR00], a model that includes both two-body and three-body interactions between nucleons. Particle fractions of neutrons, protons, electrons, muons, $\Sigma^-$ and $\Lambda$ are plotted along a logaritmic axis as a function of the total density which is given in terms of fm$^{-3}$.

used the microscopic many body calculation from [**VPR00**] over the parameterization of the Baryon-baryon potentials from [**SR99**], thus including many of the improvements that were suggested in Chapter 4. We can compare Figure 7.3 with Figure 4.5 in Chapter 4, and see that we have a good match with the bottom panel of this figure.

We see that the muon density follows the pattern of electron density, but it is lower and moves faster towards 0. The muons disappear already as the density reaches 0.49 fm$^{-3}$, while the electrons, which begin to plunge simultaneously at 0.27 fm$^{-3}$, do not disappear completely until at 0.85 fm$^{-3}$. We also see that two hyperons appear: $\Sigma^-$ at 0.27 fm$^{-3}$, and $\Lambda$ at 0.61 fm$^{-3}$. We have here thus confirmed our assumptions from Chapter 4. Although the $\Sigma^-$-hyperonet has a greater mass than $\Lambda$, it appears at lower densities. The reason for this is that the process that produces $\Sigma^-$ removes both an energetic electron and a neutron, while the disintegration of the neutral hyperon, just removes the neutron. This brings us back to the chemical potential $\mu$, which we can see in Figure 7.5. The negatively charged $\Sigma^-$-hyperonet occurs when $\mu_e + \mu_n$ corresponds to its mass, while $\Lambda$ only appears when $\mu_n$ corresponds to its mass. See otherwise Chapter 4. This will again have implications for which Urca-processes that can start in the density range under investigation.

We then look at the matter composition in Figure 7.2, where we have only included two-body interactions. We see that the point in which the electron- and muon fractions begin to sink here is at 0.35 fm$^{-3}$, which is later than in the case where we include three-body interactions. Muons also disappears later, at 0.90 fm$^{-3}$. $\Sigma^-$ begins to emerge at 0.35 fm$^{-3}$, which also lies between the values for an ideal gas and the values of the equation of state that also include three body interactions. $\Lambda$ does not apperar within our density range this time either.

We see how the different density spectra are governed by the conditions of electrical neutrality (3.26) and chemical equilibrium (3.27). When we also have muons and $\Sigma^-$ present, the equation of charge equilibrium will look like this:

$$q_p = q_e + q_\mu + q_{\Sigma^-} . \tag{7.1}$$

We thus see that the requirement of electrical neutrality, for example, makes the proton density closely follow the sum of the density of electrons, muons and $\Sigma^-$-hyperoner in Figures 7.1, 7.2 and 7.3. When we look at the chemical potentials in Figure 7.4 and Figure 7.5, kan også these can also give us hints about the number density of the different particle types. If we recapitulate Equation (3.27), it gives us the following relation between the chemical potentials of electrons, neutrons and protons in $\beta$-equilibrium:

$$\mu_n = \mu_p + \mu_e^{\text{NM}} . \tag{7.2}$$

The chemical potentials of the electrons and muons are equal, and therefore calculated from the difference between the values of the neutron and proton for this $\beta$-equilibrium requirement. This has consequences when we include the interactions in Figure 7.5. When the chemical potential of the neutron and proton eventually collapse at high densities, we see that the chemical potential of electron dives. We get the relationship between the chemical potential and number density, from the definition of the chemical potential from Equation (4.10),

$$\mu_i = \left(\frac{\partial \varepsilon}{\partial n_i}\right) , \tag{7.3}$$

and we see that the change in density is determined by the chemical potential and the change in energy for the specific particle. In our ideal gas approximation the energy will only consist of mass and kinetic energy, and we will then have a fairly simple relationship between density and energy through the chemical potential in Figure 7.4. The change in density will then develop as $\partial \varepsilon / \mu$, something we can check against the corresponding density spectrum in Figure 7.1.

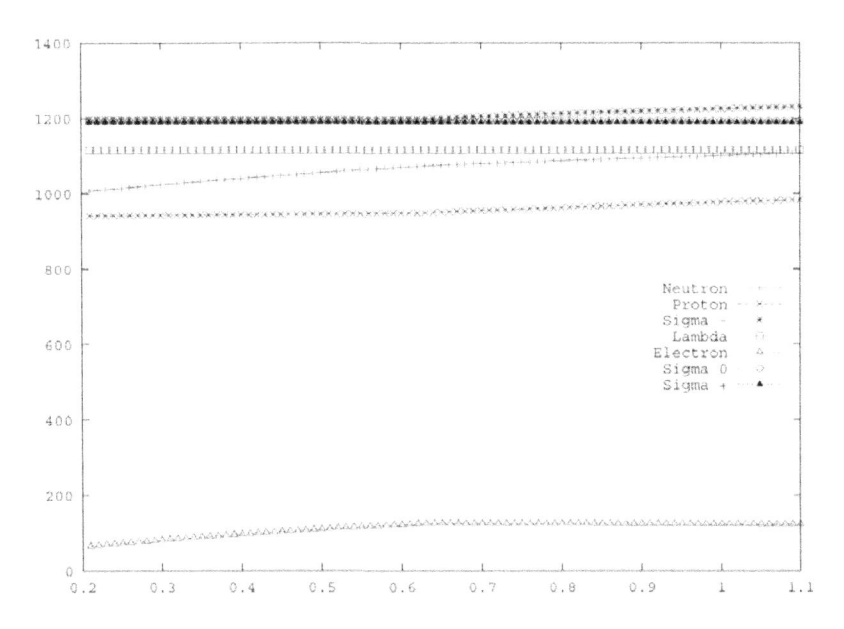

Figure 7.4: The figure shows the development of the chemical potentials of the different particle types over the density spectrum we consider in the ideal gas approximation.

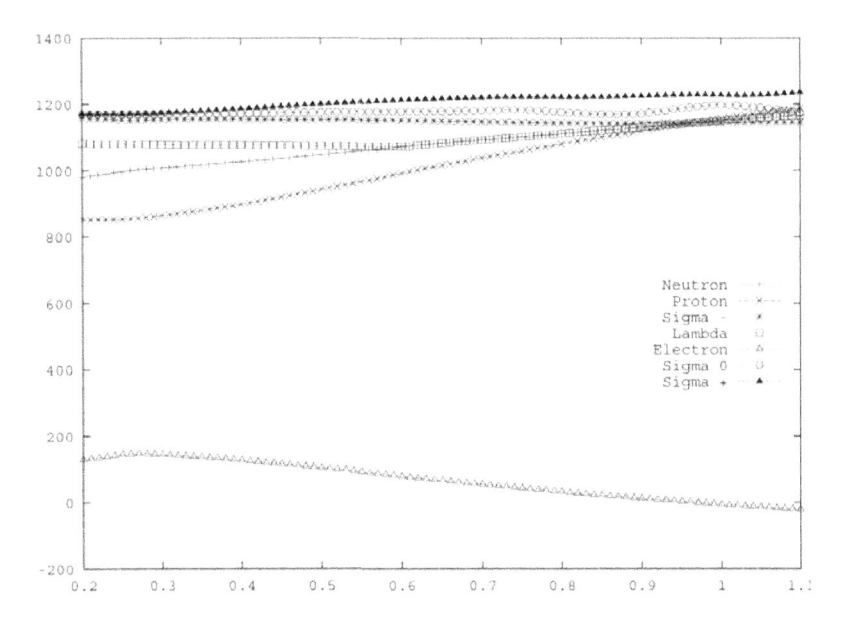

Figure 7.5: The figure shows the development of the chemical potentials of the different particle types over the density spectrum we consider for equations of state with two- and three-body interactions.

The only particle that occurs relatively equally in the density range we examine here is the neutron, although it is much more dominant at high densities before we take interactions into account. An ideal Fermi gas must thusbe said to be a bad approximation to the matter composition of the compact matter, but it can give us some general pointers about how the number density of each particle type develops with increasing density and relative to each other. When we include two-body interactions in the calculation, this improves the results considerably, but there are still significant deviations from when we include three-body interactions, so it is clear that three-body interactions play a significant role in developing a more realistic density spectrum. We also see in the expression for the emissivity in Equation (6.22) that the chemical potential of the electron plays a major role in this. We see from Figure 7.4 and 7.5 that this chemical potential shows a huge change when we take two- and three-body interactions in to account in the calculations, and this will have consequences for the neutrino emissivity. We here note that the Fermi gas model underestimates the chemical potential of electrons at the start of the density spectrum, while the opposite is the case when we come above densities of 0.5 fm$^{-3}$ (see Figure 7.4-7.5). These figures can also be compared against Figure 4.7 in Chapter 4, where the chemical potential of the electron has not been plotted, and we thus get a better representation of the other particles in the approximations.

In general, we can thus say for the composition of matter, that changes in this seems to occur at lower densities when we include interactions in to the equation of state. Hyperons such as $\Lambda$ and $\Sigma^-$ appear earlier, and electrons and muons disappear accordingly.

Now we take a look at the emissivity from the direct Urca process for the density profiles we have seen above. We calculate the emissivity for the term from Equation (6.23). For comparison of the emissivity on the figures below, the common factor $(k_B T)^6 G_F^2$ is pulled out, and the emissivity is given in units of this, where the weak coupling constant is $G_F = 8.74 \times 10^{-5}$ MeV fm$^3$. First we take a look at the neutrino emissivity of the various processes in our ideal gas approximation. We see here that in our density range, only two processes are running; $\Sigma^- \to n + e + \bar{\nu}_e$ and $\Sigma^- \to n + \mu + \bar{\nu}_\mu$. Both processes start simultaneously, immediately after the $\Sigma^-$ appear at the density range of 0.63 fm$^{-3}$. These processes will both will be equally powerful as long as they run, as the chemical potentials of electrons and muons are equal. Here both run to the end of the density range under investigation. Similarly, we see from Figure 7.1 that the density of muons and electrons continue beyond the density range.

When we look at the emissivity of our approximation with only two-body interactions in Figure 7.7, we see that it is similar to that for an ideal gas,

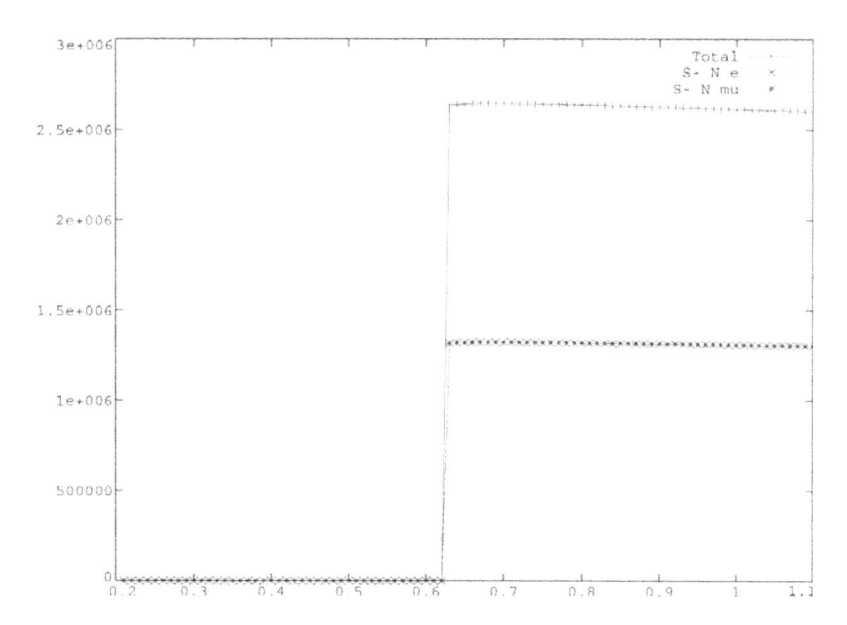

[Figure: plot of neutrino emissivity vs total density, with curves labeled "Total", "S- N e", and "S- N mu"]

Figure 7.6: Here we show the neutrino emissivity from the two processes running in the ideal gas approximation as a function of the total density. $\Sigma^- \to n + e + \bar{\nu}_e$ is marked as "S- N e", and $\Sigma^- \to n + \mu + \bar{\nu}_\mu$ is marked as "S- N mu".

except that the $\Sigma^-$ process starts earlier, at a density of 0.36 fm$^{-3}$, whereas when we take three-body interactions into account they start already at 0.28 fm$^{-3}$. If we look at the composition of matter in this case, we see that the density of muons drops to 0 at 0.9 fm$^{-3}$, making the process with muons naturally enough come to a halt.

When we look at the emissivity of the density spectrum in Figure 7.8 where we also have included three-body interactions, we see that the processes $\Sigma^- \to n + e + \bar{\nu}_e$ and $\Sigma^- \to n + \mu + \bar{\nu}_\mu$ appear simultaneously at 0.28 fm$^{-3}$. These begin immediately when the $\Sigma^-$-hyperon appears in the nuclear matter (see Figure 7.3), which corresponds to when the fraction of electrons and muons begins to fall. These processes will be equally powerful here as well, as the chemical potentials of the electron and the muon are equal. We see how these processes run together until the muon fraction becomes too small, and the process with muons stop at 0.50 fm$^{-3}$.

We thus see that although it may seem that the role the three-body interaction plays is not as great as the importance of the two-body interaction for the particle composition, it is of critical importance for the neutrino emissions in the density range under investigation. Since the $\Lambda$-hyperon does not appear in this density range unless we take the three-body interaction into account, we obviously neither get the Urca process with the $\Lambda$-hyperon

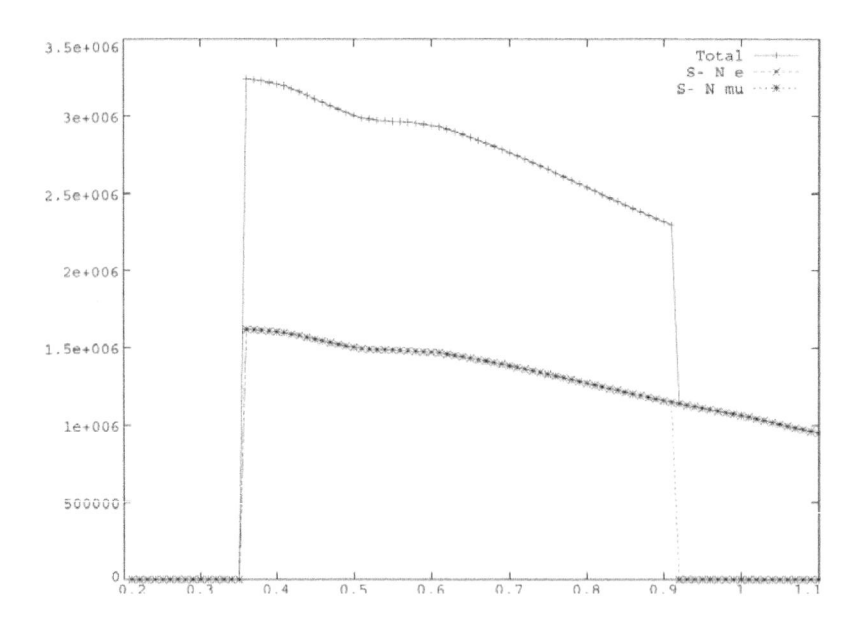

Figure 7.7: Here we show the neutrino emissivity of the Direct Urca processes which appear under the equation of state with only two-body interactions, plotted as a function of the total density: $\Sigma^- \to n + e + \bar{\nu}_e$, marked as "S- N e", $\Sigma^- \to n + \mu + \bar{\nu}_\mu$, marked as "S- N mu" og $\Lambda \to p + e + \bar{\nu}_e$, marked as "L P e".

to run without it. Similarly, we also see how the electron and the muon here disappear much earlier.

Another major difference from our ideal gas approximation is the development of the emissivity, how it in our approximation with interactions in Figure 7.8, as soon as it appears, begins to fall, in contrast to what we get with an ideal gas approximation. If we look at the expression for the emissivity in Equation (6.22), and also look at the change in the chemical potential between Figure 7.4 and 7.5, we will immediately see that the chemical potential of the electron must take much of the blame for this. At the time the direct Urca process starts in the ideal gas approximation, the electrons chemical potential is already somewhat lower when we include three-body interactions, and it decreases further. Thus we see how the chemical potential is crucial for the emissivity and how it changes significantly when we include two-and three-body interactions.

Finally, we briefly discuss what changes we expect if calculations are done for the more detailed model for neutrino emissions that we introduced in the second part of Chapter 6. If we look at Equation (6.57)-(6.61), we see that the Fermi level here will play a central role in the changes we get. Such a calculation requires that we know the chemical potentials of the different particle types that are included in the reaction. We have an approximation

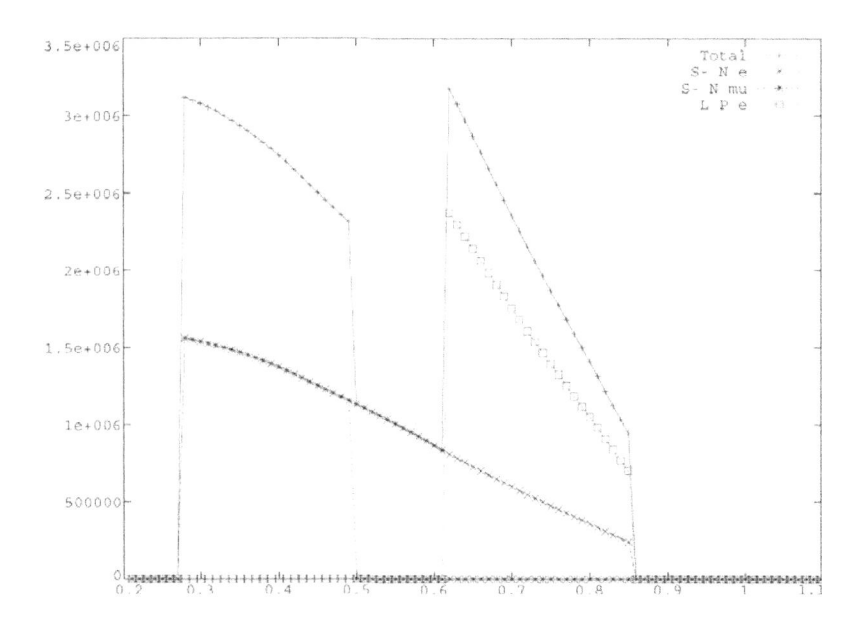

Figure 7.8: Here we show the neutrino emissivity of the Direct Urca processes which appear under the equation of state from [VPR00] with both two-body and three-body interactions: $\Sigma^- \to n+e+\bar{\nu}_e$, marked as "S- N e ", $\Sigma^- \to n+\mu+\bar{\nu}_\mu$, marked as "S- N mu" og $\Lambda \to p+e+\bar{\nu}_e$, marked as "L P e".

to these from a many body calculation, but there are sufficiently many uncertainties surrounding these, that a more exact calculation will not necessarily result in drastic improvements. You will still be able to expect somewhat more accurate results, but within the same order of magnitude as those we have estimated earlier in this chapter.

Neutrino emissions will thus be the main source of cooling of a neutron star in the earliest stages of its life, and it is assumed that the direct Urca process is the most powerful source of energy loss during this time. Nevertheless, the modified Urca process which was discussed in Chapter 5 should be dominant in the outer core and in the cores of neutron stars with relatively smaller masses. The modified Urca processes can run at smaller densities as they are helped by a companion particle to conserve the momentum and chemical potential in the process. We will now compare the rate of the direct Urca process with the modified Urca process in Equation (5.11)-(5.14). The modified Urca process can be viewed as a correction to the direct Urca process because of the damping of the contributing particles from collisions. From Frimann and Maxwell [**FM79**] we have, for the neutrino emissivity from the modified Urca process of free particles, the expression

$$\varepsilon^{MOD} = \frac{11513}{60480} \frac{G^2 g_1^2 m_n^3 m_p}{2\pi\hbar} \left(\frac{f}{m_\pi}\right)^4 p_F(e) \, \alpha_{URCA} \, (kT)^8 \quad , \qquad (7.4)$$

where $\alpha_{URCA}$ is a function that depends on the masses and Fermi levels the contributing particles. For a rough comparison with the direct Urca process we set this approximately equal to the value at the saturation density, $\alpha_{URCA} \approx 1.13$. $f$ is the p-wave $\pi N$ coupling constant, which we like Frimann and Maxwell put $\approx 1$ in this comparison. Similarly, we write the factor $p_F(e) \approx 85 \left(\rho/\rho_0\right)^{\frac{2}{3}} \frac{\text{Mev}}{c}$, and we set the saturation density $\rho_0 = 0.16 \text{ fm}^{-3}$.

When we then make a dimensional analysis, and compare this with the direct Urca process from Equation (6.23), we see that Frimann and Maxwell [**FM79**] in the calculation of the expression for the emissivity at times has set $\hbar c = 1$, and at times $\hbar = c = 1$, which means that they end up with only one $\hbar$ in the expression in Equation (7.4). This in contrast to Prakash et al [**LPP91, PPL92**] who include these constants in their expression in Equation (6.23). Furthermore, we have $G$ given by MeV fm$^3$, while the factor $p_F(e)$ is given by MeV/$c$, ie. MeV s m$^{-1}$. $kT$ gives us the total MeV, while the different masses give us MeV/$c^2 \to$ MeV s$^2$ m$^{-2}$. When we insert for these, we saw in Chapter 6 that in the expression from [**LPP91**] we got the unit MeV s$^{-1}$ m$^{-3}$, while we in the expression from [**FM79**] divide by $c^8 \hbar^9$ to get the same units, so that the emissivity is given by a unit of energy per time per volume. For comparison, we use cgs units, and we make a small table where we give the relationship between the emissivity for different densities and temperatures. Since we are here to look at the relationship between the two processes, we introduce the fraction

$$\frac{\varepsilon^{MOD}}{\varepsilon^{DIR}} = \frac{11513 \cdot 10080 \cdot 85 m_n^2 \left(\frac{\rho}{\rho_0}\right)^{\frac{2}{3}} (kT)^2}{457 \cdot 60480 \cdot 2\pi^2 m_\pi^4 \mu_e c^3} \quad , \tag{7.5}$$

where we multiply and contract together the constants and get

$$\frac{\varepsilon^{mod}}{\varepsilon^{dir}} \approx 1,37 \cdot 10^{-21} \frac{\rho^{2/3}}{\mu_e} T^2 \quad , \tag{7.6}$$

that we use for the calculation in Table 7.1 below, together with the estimated values for the chemical potential $\mu_e$.

Table 7.1

Comparison between the neutrino emissivity from the direct and modified Urca process

| $\rho$ | $\mu_e$ | $T$ | $\varepsilon^{mod}/\varepsilon^{dir}$ |
|---|---|---|---|
| 0.3 fm$^{-3}$ | 144.7 MeV | $10^8$ K | $4.24 \cdot 10^{-8}$ |
| 0.3 fm$^{-3}$ | 144.7 MeV | $10^9$ K | $4.24 \cdot 10^{-6}$ |
| 0.5 fm$^{-3}$ | 104.9 MeV | $10^8$ K | $8.22 \cdot 10^{-8}$ |
| 0.5 fm$^{-3}$ | 104.9 MeV | $10^9$ K | $8.22 \cdot 10^{-6}$ |
| 0.6 fm$^{-3}$ | 79.3 MeV | $10^8$ K | $1.23 \cdot 10^{-7}$ |
| 0.6 fm$^{-3}$ | 79.3 MeV | $10^9$ K | $1.23 \cdot 10^{-5}$ |

We now see that the direct Urca process dominates during temperatures of $10^8$ - $10^9$ K. We have to increase to temperatures of $T \simeq T_F \simeq 10^{12}$ K for the modified Urca process to be of the same order of magnitude. If we roughly set $\frac{\rho^{2/3}}{\mu_e} \sim 0,01$ and insert $T_9$ for $T$, we get $\varepsilon^{mod}/\varepsilon^{dir} \sim 10^{-5} T_9^2$, which corresponds to results from other comparisons between the processes, for example in [**PPL92**]. This is close to equivalent to a factor of $(T/T_F)^2$, which we can interpret as that the companion particles under the modified Urca process in both the primordial and final state provide a factor $T/T_F$. If we then compare this with the processes we can get to run with our equation of state, we see that the variables that are different in the expression for the emissivity, particle masses and constants in Table 6.1, will give the results of the same order of magnitude. The modified Urca process is thus of little significance, given that the direct Urca-process is running. Nevertheless, it is important as it may run at lower densities and thus may be responsible for a significant part of the neutrino emissivity, and thus also the cooling we have of the star when the direct Urca process is not running.

# Chapter 8

# Conclusion

In this short booklet we have given a general introduction to the physics of neutron stars, and especially the neutrino emissions from these. Neutron stars are the end point in the evolution of stars with masses in the range between Chandrasekhar mass (ca. 1.4 $M_\odot$) and the critical mass for the formation of black holes (ca. 3-5 $M_\odot$). Furthermore, we have looked at the physical theory behind the development of different equations of state and the consequences the choice of equation of state has is of the properties, including the mass and radius we calculate for neutron stars. We then considered the inclusion of hyperons, what consequences this has for the equation of state and the interactions we must take into account to provide a good description of densities and the equation of state of a neutron star with hyperons. We then seek to tie all this up towards the cooling of neutron stars, and then specifically the issue of neutrinos, who are an important part of this cooling process. Neutrino emissions dominate the cooling of the neutron star from the formation and until it reaches a temperature of $\sim 10^8$ K. The neutrino processes thus dominate the coolong over a period of perhaps 100,000 years, and then typically in a temperature range from $10^9$-$10^8$ K for the inner temperature.

The most effective and strongest of these neutrino processes is the direct Urca process, described in Equation (5.25)-(5.26). More accurate calculations of the emissivity of the direct Urca on the model of Equation (6.57)-(6.61), will probably be able to give a somewhat, but not dramatically better estimate of the emissivity of this process. When the direct Urca process is running, it will be dominant over other processes such as the modified Urca process. In Chapter 7 we saw that when both processes run at a temperature of $\sim 10^9$ K, the direct Urca dominate by a factor $\sim 10^5$.

Furthermore, we looked at the processes would run with our various iterations of the equations of state. With the ideal gas model for the equation of state, we can only get the process $\Sigma^- \to n+l+\bar{\nu}_l$ to run, with both electrons and muons. We get a similar result when we include two-body interactions. The change here is mainly that the process starts earlier and declines faster. The inclusion of two-body interactions thus has a significant, but not dramatic effect on the emissivity. When we include three-body interactions we see, however, a marked difference. Both the size of the emissivity and especially the shape of the graphs changed dramatically due to changes in the electron chemical potential. Further, changes in the density spectra allows the process $\Lambda \to p+e+\bar{\nu}_e$ to run, something that gives us neutrino emissions at higher densities than for the process with the $\Sigma^-$ hyperon. In general we can say that without the inclusion of two-and three-body interactions we get a poor a poor approximation for the neutrino emissivity from the direct Urca process.

We also see that with the matter composition from the equation of state we have selected in Figure 7.3, we do not get the direct Urca process with nucleons running. If we look at the Fermi levels we get out of the program in Appendix A, we see that the conditions of the triangle inequality, which constitutes the $\theta$-function in Equation (6.23), are not met in the density range we are examining. We nevertheless see that the margins are narrow, and one could perhaps imagine that the process can run with only minor adjustments to the equation of state. If so, this could be a significant contributor to the neutrino emissivity. If this process does not run, the direct Urca process with hyperons will be an even more important contributor to the cooling of neutron stars.

The conclusion of this summary must therefore be that when you have hyperons present in the matter, to get a good estimate of the neutrino emissivity from a neutron star, one has to tak into account the possibility of Urca processes with hyperon matter and two-and three-body interactions in the calculations.

# Appendix A

# Program for calculating the emissivity

```
!*****************************************
!Program Urcaemissivities.f90
! R. Kjelsberg 2003
! The program calculates neutrino emissivities from
! the direct Urca process.
!
!*****************************************
! This module contains the subroutines to
! manipulate the indata to the program.
! Allocate, read, write, convert, calculate
!
MODULE particle_density
TYPE, PUBLIC :: general_array
INTEGER :: n_data
DOUBLE PRECISION, DIMENSION(:), POINTER :: &
total_baryon, neutron, &
proton, electron, muon, sigma_minus,&
lambda, sigma0, sigma_plus, &
epsilon_nu_npe,epsilon_nu_npm,&
epsilon_nu_lpe,&
epsilon_nu_lpm,epsilon_nu_smne,&
epsilon_nu_smnm,epsilon_nu_smle,&
epsilon_nu_smlm,epsilon_nu_sms0e,&
epsilon_nu_sms0m, epsilon_tot1
END TYPE general_array
TYPE (general_array), &
```

```fortran
PUBLIC :: density_distribution, fermilevels_array, &
energies_array, chem_array, epsilon_array
CONTAINS
!****************************************************
!This subroutine allocates room for densities, fermilevels and energies
!
SUBROUTINE allocate_dd_array(this_array)
IMPLICIT NONE
INTEGER :: i
TYPE(general_array), INTENT(INOUT) :: this_array
INTEGER :: n
n = this_array%n_data
IF (ASSOCIATED(this_array%total_baryon))&
DEALLOCATE(this_array%total_baryon)
IF (ASSOCIATED(this_array%neutron)) &
DEALLOCATE(this_array%neutron)
IF (ASSOCIATED(this_array%proton))&
DEALLOCATE(this_array%proton)
IF (ASSOCIATED(this_array%electron))&
DEALLOCATE(this_array%electron)
IF (ASSOCIATED(this_array%muon))&
DEALLOCATE(this_array%muon)
IF (ASSOCIATED(this_array%sigma_minus))&
DEALLOCATE(this_array%sigma_minus)
IF (ASSOCIATED(this_array%lambda))&
DEALLOCATE(this_array%lambda)
IF (ASSOCIATED(this_array%sigma0))&
DEALLOCATE(this_array%sigma0)
IF (ASSOCIATED(this_array%sigma_plus))&
DEALLOCATE(this_array%sigma_plus)
! This allocates room for the particle densities
ALLOCATE(this_array%total_baryon(n))
ALLOCATE(this_array%neutron(n))
ALLOCATE(this_array%proton(n))
ALLOCATE(this_array%electron(n))
ALLOCATE(this_array%muon(n))
ALLOCATE(this_array%sigma_minus(n))
ALLOCATE(this_array%lambda(n))
ALLOCATE(this_array%sigma0(n))
ALLOCATE(this_array%sigma_plus(n))
DO i = 1,this_array%n_data
```

```fortran
      this_array%total_baryon(i) = 0
      this_array%neutron(i) = 0
      this_array%proton(i)=0
      this_array%electron(i)=0
      this_array%muon(i)=0
      this_array%sigma_minus(i)=0
      this_array%lambda(i)=0
      this_array%sigma0(i)=0
      this_array%sigma_plus(i)=0
      ENDDO
      END SUBROUTINE allocate_dd_array
!***********************************************************
! This subroutine reads the data from the dens.dat file to
! a new set of variables
!
      SUBROUTINE read_dd_data(this_array)
      IMPLICIT NONE
      INTEGER :: i
      TYPE(general_array), INTENT(INOUT) :: this_array
      CHARACTER (LEN = 100), POINTER ::densities
      DOUBLE PRECISION :: tot1, ne1, p1, e_min1,&
      mu_min1, sigma_min1, &
      lambda1, sigma_null1
      DO i=1, this_array%n_data
      READ(1,*)tot1, ne1, p1, e_min1,&
      mu_min1, sigma_min1, lambda1,&
      sigma_null1, sigma_pluss1
      this_array%total_baryon(i)=tot1
      this_array%neutron(i)=ne1
      this_array%proton(i)=p1
      this_array%electron(i)=e_min1
      this_array%muon(i)=mu_min1
      this_array%sigma_minus(i)=sigma_min1
      this_array%lambda(i)=lambda1
      this_array%sigma0(i)=sigma_null1
      this_array%sigma_plus(i)=sigma_pluss1
      ENDDO
      END SUBROUTINE read_dd_data
!***********************************************************
!This subroutine writes the data from the dens.dat file
! to a new file densout.dat
```

```fortran
!
SUBROUTINE write_dd_data(this_array)
IMPLICIT NONE
INTEGER :: i
TYPE(general_array), INTENT(IN) :: this_array
! CHARACTER (LEN = 100), POINTER :: densities
DOUBLE PRECISION :: tot1, ne1, p1, e_min1,&
mu_min1, sigma_min1, &
lambda1, sigma_null1, sigma_pluss1
OPEN(UNIT=2, FILE='densout.dat')
DO i= 1,this_array%n_data
WRITE(2,'(9(d12.6,2x))') this_array%total_baryon(i),&
this_array%neutron(i), &
this_array%proton(i),this_array%electron(i), &
this_array%muon(i), this_array%sigma_minus(i),&
this_array%lambda(i),this_array%sigma0(i),&
this_array%sigma_plus(i)
ENDDO
END SUBROUTINE write_dd_data
!***********************************************************
! This subroutine converts the densities to find the fermilevels.
!
SUBROUTINE fermilevel_convertion
IMPLICIT NONE
INTEGER :: i
! TYPE(general_array)this_array
DOUBLE PRECISION :: pi2
pi2 = acos(-1.0d0)**2.0d0
OPEN(UNIT=5,FILE="fermilevels.dat")
DO i = 1,fermilevels_array%n_data
fermilevels_array%neutron(i) = &
(density_distribution%neutron(i)*3.*pi2)**(1./3.)
fermilevels_array%proton(i) = &
(density_distribution%proton(i)*3.*pi2)**(1./3.)
fermilevels_array%electron(i) = &
(density_distribution%electron(i)*3.*pi2)**(1./3.)
fermilevels_array%muon(i) = &
(density_distribution%muon(i)*3.*pi2)**(1./3.)
fermilevels_array%sigma_minus(i) = (&
density_distribution%sigma_minus(i)*3.*pi2)**(1./3.)
fermilevels_array%lambda(i) = &
```

```fortran
(density_distribution%lambda(i)*3.*pi2)**(1./3.)
fermilevels_array%sigma0(i) = &
(density_distribution%sigma0(i)*3.*pi2)**(1./3.)
fermilevels_array%sigma_plus(i) = &
(density_distribution%sigma_plus(i)*3.*pi2)**(1./3.)
WRITE (5,'(8(d12.6,2x))')fermilevels_array%neutron(i),&
fermilevels_array%proton(i),fermilevels_array%electron(i),&
fermilevels_array%muon(i),fermilevels_array%sigma_minus(i),&
fermilevels_array%lambda(i),&
fermilevels_array%sigma0(i),fermilevels_array%sigma_plus(i)
ENDDO
END SUBROUTINE fermilevel_convertion
!***********************************************************
! Now we must read from the chem.dat file to
! get the value for the electron/muon we need in the calculations
! to find the emissivity.
!
SUBROUTINE read_chem_data(this_array)
IMPLICIT NONE
INTEGER :: i
TYPE(general_array), INTENT(INOUT) :: this_array
DOUBLE PRECISION :: chemdens, ne1chem, &
p1chem, e_min1chem, sigma_min1chem, &
lambda1chem, sigma_null1chem, sigma_pluss1chem
OPEN(UNIT=3, FILE='chem.dat')
DO i=1, this_array%n_data
READ(3,*)chemdens, ne1chem, p1chem, &
e_min1chem, sigma_min1chem, &
lambda1chem, sigma_null1chem, sigma_pluss1chem
this_array%neutron(i)=ne1chem
this_array%proton(i)=p1chem
this_array%electron(i)=e_min1chem
this_array%sigma_minus(i)=sigma_min1chem
this_array%lambda(i)=lambda1chem
this_array%sigma0(i)=sigma_null1chem
this_array%sigma_plus(i)=sigma_pluss1chem
ENDDO
END SUBROUTINE read_chem_data
!*********************************************************
!This subroutine allocates room for the emissivities.
!
```

```fortran
SUBROUTINE allocate_eps_array(this_array)
IMPLICIT NONE
TYPE(general_array), INTENT(INOUT) :: this_array
INTEGER :: n, i
n = this_array%n_data
IF (ASSOCIATED(this_array%epsilon_nu_npe)) &
DEALLOCATE(this_array%epsilon_nu_npe)
IF (ASSOCIATED(this_array%epsilon_nu_npm)) &
DEALLOCATE(this_array%epsilon_nu_npm)
IF (ASSOCIATED(this_array%epsilon_nu_lpe)) &
DEALLOCATE(this_array%epsilon_nu_lpe)
IF (ASSOCIATED(this_array%epsilon_nu_lpm)) &
DEALLOCATE(this_array%epsilon_nu_lpm)
IF (ASSOCIATED(this_array%epsilon_nu_smne)) &
DEALLOCATE(this_array%epsilon_nu_smne)
IF (ASSOCIATED(this_array%epsilon_nu_smnm)) &
DEALLOCATE(this_array%epsilon_nu_smnm)
IF (ASSOCIATED(this_array%epsilon_nu_smle)) &
DEALLOCATE(this_array%epsilon_nu_smle)
IF (ASSOCIATED(this_array%epsilon_nu_smlm)) &
DEALLOCATE(this_array%epsilon_nu_smlm)
IF (ASSOCIATED(this_array%epsilon_nu_sms0e)) &
DEALLOCATE(this_array%epsilon_nu_sms0e)
IF (ASSOCIATED(this_array%epsilon_nu_sms0m)) &
DEALLOCATE(this_array%epsilon_nu_sms0m)
IF (ASSOCIATED(this_array%epsilon_tot1)) &
DEALLOCATE(this_array%epsilon_tot1)
! This allocates room for the particle densities
ALLOCATE(this_array%epsilon_nu_npe(n))
ALLOCATE(this_array%epsilon_nu_npm(n))
ALLOCATE(this_array%epsilon_nu_lpe(n))
ALLOCATE(this_array%epsilon_nu_lpm(n))
ALLOCATE(this_array%epsilon_nu_smne(n))
ALLOCATE(this_array%epsilon_nu_smnm(n))
ALLOCATE(this_array%epsilon_nu_smle(n))
ALLOCATE(this_array%epsilon_nu_smlm(n))
ALLOCATE(this_array%epsilon_nu_sms0e(n))
ALLOCATE(this_array%epsilon_nu_sms0m(n))
ALLOCATE(this_array%epsilon_tot1(n))
DO i = 1,this_array%n_data
this_array%epsilon_nu_npe(i) = 0
```

```fortran
this_array%epsilon_nu_npm(i) = 0
this_array%epsilon_nu_lpe(i)=0
this_array%epsilon_nu_lpm(i)=0
this_array%epsilon_nu_smne(i)=0
this_array%epsilon_nu_smnm(i)=0
this_array%epsilon_nu_smle(i)=0
this_array%epsilon_nu_smlm(i)=0
this_array%epsilon_nu_sms0e(i)=0
this_array%epsilon_nu_sms0m(i)=0
this_array%epsilon_tot1(i)=0
ENDDO
END SUBROUTINE allocate_eps_array
!***********************************************************
! This subroutine tests the triangle inequality
! to see which processes can go, and
! calculate the neutrino emissivity epsilon_nu
! by the formula from Prakash et.al.
! ( Phys.Rev.Lett 66 (1991) 2703)
!
! All results are given in units of
! [[G_F**2]*[<(k_B*T)**6>],
! as we omit these constants from our calculations.
! Masses are in MeV/c**2
!
SUBROUTINE test_fermilevel
IMPLICIT NONE
DOUBLE PRECISION::root23, step
INTEGER :: i
root23 = (2.0d0/3.0d0)**(1.0d0/2.0d0)
DO i = 1, epsilon_array%n_data
! First we test the process n > p e v
IF (fermilevels_array%electron(i)>0&
.AND. fermilevels_array%proton(i)>0&
.AND. fermilevels_array%neutron(i) > 0&
.AND. fermilevels_array%electron(i)&
+ fermilevels_array%proton(i) - &
fermilevels_array%neutron(i) > 0) THEN
epsilon_array%epsilon_nu_npe(i) = (0.142431334)*&
((0.97390499)**2.0d0 * &
(1.0d0+&
3.0d0*1.233**2.0d0)*&
```

```
          939.56563*&
          938.27231 *&
          (chem_array%electron(i)))
     ENDIF
     ! Then the process n > p mu v
     IF (fermilevels_array%muon(i)>0&
     .AND. fermilevels_array%proton(i)>0&
     .AND. fermilevels_array%neutron(i) > 0&
     .AND. fermilevels_array%muon(i) &
     + fermilevels_array%proton(i) - &
     fermilevels_array%neutron(i) > 0) THEN
          epsilon_array%epsilon_nu_npm(i) = (0.142431334)*&
          ((0.97390499)**2.0d0 * &
          (1.0d0+&
          3.0d0*1.233**2.0d0)&
          *939.56563&
          *938.27231 *&
          (chem_array%electron(i)))
     ENDIF
     ! Then the process lambda > p e v
     IF (fermilevels_array%electron(i)>0&
     .AND. fermilevels_array%proton(i)>0&
     .AND. fermilevels_array%lambda(i) > 0&
     .AND. fermilevels_array%electron(i) &
     + fermilevels_array%proton(i) - &
     fermilevels_array%lambda(i) > 0) THEN
          epsilon_array%epsilon_nu_lpe(i) = (0.142431334)*&
          ((0.231)**2.0d0 * &
          (1.50d0+&
          3*1.5*(0.477+0.756/3)**2.0d0)&
          *1115.684&
          *938.27231 &
          *(chem_array%electron(i)))
     ENDIF
     ! Then the process lambda > p mu v
     IF (fermilevels_array%muon(i)>0&
     .AND. fermilevels_array%proton(i)>0&
     .AND. fermilevels_array%lambda(i) > 0&
     .AND. fermilevels_array%muon(i)&
     + fermilevels_array%proton(i) - &
     fermilevels_array%lambda(i) > 0) THEN
```

```
epsilon_array%epsilon_nu_lpm(i)= (0.142431334)*&
((0.231)**2.0d0 * &
(1.50d0+&
4.5*(0.477+0.756/3)**2.0d0)&
*1115.684&
*938.27231 &
*(chem_array%electron(i)))
ENDIF
! Then the process sigma_minus > n e v
IF (fermilevels_array%electron(i)>0&
.AND. fermilevels_array%neutron(i)>0&
.AND. fermilevels_array%sigma_minus(i) > 0&
.AND. fermilevels_array%electron(i)&
+ fermilevels_array%neutron(i) - &
fermilevels_array%sigma_minus(i) > 0) THEN
epsilon_array%epsilon_nu_smne(i) = (0.142431334)*&
((0.231)**2.0d0 * &
(1.0d0+&
3.0d0*(0.756-0.477)**2.0d0)&
*1197.436&
*939.56563 &
*(chem_array%electron(i)))
ENDIF
! Then the process sigma_minus > n mu v
IF (fermilevels_array%muon(i)>0&
.AND. fermilevels_array%neutron(i)>0&
.AND. fermilevels_array%sigma_minus(i) > 0&
.AND. fermilevels_array%muon(i)&
+ fermilevels_array%neutron(i) - &
fermilevels_array%sigma_minus(i) > 0) THEN
epsilon_array%epsilon_nu_smnm(i) = (0.142431334)*&
((0.231)**2.0d0 * &
(1.0d0+&
3.0d0*(0.756-0.477)**2.0d0)&
*1197.436 &
*939.56563 &
*(chem_array%electron(i)))
ENDIF
! Then the process sigma_minus > lambda e v
IF (fermilevels_array%electron(i)>0&
.AND. fermilevels_array%lambda(i)>0&
```

```fortran
.AND. fermilevels_array%sigma_minus(i) > 0&
.AND. fermilevels_array%electron(i)&
+ fermilevels_array%lambda(i) - &
fermilevels_array%sigma_minus(i) > 0) THEN
epsilon_array%epsilon_nu_smle(i) = (0.142431334)*&
((0.97390499)**2.0d0 * &
(1.0d0+2.0d0*(0.756)**2.0d0)&
*1197.436&
*1115.684 &
*(chem_array%electron(i)))
ENDIF
! Then the process sigma_minus > lambda mu v
IF (fermilevels_array%muon(i)>0&
.AND. fermilevels_array%lambda(i)>0&
.AND. fermilevels_array%sigma_minus(i) > 0&
.AND. fermilevels_array%muon(i)&
+ fermilevels_array%lambda(i) - &
fermilevels_array%sigma_minus(i) > 0) THEN
epsilon_array%epsilon_nu_smlm(i) = (0.142431334)*&
((0.97390499)**2.0d0 * &
(1.0d0+2.0d0*(0.756)**2.0d0)&
*1197.436&
*1115.684 &
*(chem_array%electron(i)))
ENDIF
! Then the process sigma_minus > sigma0 e v
IF (fermilevels_array%electron(i) > 0&
.AND. fermilevels_array%sigma0(i) > 0&
.AND. fermilevels_array%sigma_minus(i) > 0&
.AND. fermilevels_array%electron(i) &
+ fermilevels_array%sigma0(i) - &
fermilevels_array%sigma_minus(i) > 0) THEN
epsilon_array%epsilon_nu_sms0e(i) = (0.142431334)*&
((0.97390499)**2.0d0 * &
(1.0d0+&
6.0d0*(0.477)**2.0d0)&
*1197.436&
*1192.55 &
*(chem_array%electron(i)))
ENDIF
! Then the process sigma_minus > sigma0 mu v
```

```fortran
IF (fermilevels_array%muon(i) > 0&
.AND. fermilevels_array%sigma0(i) > 0&
.AND. fermilevels_array%sigma_minus(i) > 0&
.AND. fermilevels_array%muon(i)&
+ fermilevels_array%sigma0(i) - &
fermilevels_array%sigma_minus(i) > 0) THEN
epsilon_array%epsilon_nu_sms0m(i) = (0.142431334)*&
((0.97390499)**2.0d0 * &
(2.0d0+&
6.0d0*(0.477)**2.0d0)&
*1197.436&
*1192.55 &
*(chem_array%electron(i)))
ENDIF
! Now to calculate the total emissivity from this estimate.
epsilon_array%epsilon_tot1(i) = epsilon_array%epsilon_nu_npe(i) +&
epsilon_array% epsilon_nu_npm(i) + &
epsilon_array%epsilon_nu_lpe(i) + &
epsilon_array%epsilon_nu_lpm(i) + &
epsilon_array%epsilon_nu_smne(i) + &
epsilon_array%epsilon_nu_smnm(i) +&
epsilon_array%epsilon_nu_smle(i) +&
epsilon_array%epsilon_nu_smlm(i) +&
epsilon_array%epsilon_nu_sms0e(i) +&
epsilon_array%epsilon_nu_sms0m(i)
WRITE(4,'(11(d12.6,2x))') epsilon_array%epsilon_nu_npe(i) , &
epsilon_array% epsilon_nu_npm(i), &
epsilon_array%epsilon_nu_lpe(i), &
epsilon_array%epsilon_nu_lpm(i), &
epsilon_array%epsilon_nu_smne(i), &
epsilon_array%epsilon_nu_smnm(i), &
epsilon_array%epsilon_nu_smle(i), &
epsilon_array%epsilon_nu_smlm(i), &
epsilon_array%epsilon_nu_sms0e(i), &
epsilon_array%epsilon_nu_sms0m(i), &
epsilon_array%epsilon_tot1(i)
! The parameter "step" simulates the steps of increasing density
step= .2+0.01*i
WRITE(*,'(12(d12.6,2x))')step, &
epsilon_array%epsilon_nu_npe(i) , &
epsilon_array%epsilon_nu_npm(i) , &
```

```fortran
                    epsilon_array%epsilon_nu_lpe(i), &
                    epsilon_array%epsilon_nu_lpm(i), &
                    epsilon_array%epsilon_nu_smne(i), &
                    epsilon_array%epsilon_nu_smnm(i), &
                    epsilon_array%epsilon_nu_smle(i), &
                    epsilon_array%epsilon_nu_smlm(i), &
                    epsilon_array%epsilon_nu_sms0e(i), &
                    epsilon_array%epsilon_nu_sms0m(i), &
                    epsilon_array%epsilon_tot1(i)
    WRITE(7,'(12(d12.6,2x))')step, &
                    epsilon_array%epsilon_nu_npe(i) , &
                    epsilon_array%epsilon_nu_npm(i) , &
                    epsilon_array%epsilon_nu_lpe(i), &
                    epsilon_array%epsilon_nu_lpm(i), &
                    epsilon_array%epsilon_nu_smne(i), &
                    epsilon_array%epsilon_nu_smnm(i), &
                    epsilon_array%epsilon_nu_smle(i), &
                    epsilon_array%epsilon_nu_smlm(i), &
                    epsilon_array%epsilon_nu_sms0e(i), &
                    epsilon_array%epsilon_nu_sms0m(i), &
                    epsilon_array%epsilon_tot1(i)
    ENDDO
    END SUBROUTINE test_fermilevel
    END MODULE particle_density
!*********************************************
! We now start the main program.
!
PROGRAM prog
USE particle_density
IMPLICIT NONE
INTEGER :: n
OPEN(UNIT=1,FILE="dens.dat")
OPEN(UNIT=4,FILE="epsilon1.dat")
OPEN(UNIT=7,FILE="plot11.dat")
READ(1,*) n
density_distribution%n_data = n
fermilevels_array%n_data = n
energies_array%n_data = n
epsilon_array%n_data = n
chem_array%n_data = n
CALL allocate_dd_array(density_distribution)
```

```
CALL allocate_dd_array(fermilevels_array)
CALL allocate_dd_array(energies_array)
CALL allocate_dd_array(chem_array)
CALL allocate_eps_array(epsilon_array)
CALL read_dd_data(density_distribution)
CALL read_chem_data(chem_array)
CALL write_dd_data(density_distribution)
CALL fermilevel_convertion
CALL test_fermilevel
END PROGRAM prog
```

# Appendix B

# Calculations in Maple

>     int(1/(1+exp(-x)))*1/(1+exp(x+v)),x)

$$\frac{\ln\left(1+e^{x}e^{v}\right)}{e^{v}-1}-\frac{\ln\left(e^{x}+1\right)}{e^{v}-1} \tag{B.1}$$

>     int(1/(1+exp(x)))*1/(1+exp(-x+v)),x)

$$-\frac{\ln\left(e^{x}+e^{v}\right)}{e^{v}-1}+\frac{\ln\left(e^{x}+1\right)}{e^{v}-1} \tag{B.2}$$

>     int((Pi^2)*y^3/(2*(exp(y)+1)),y=0..infinity)
            +((y^5/(2*(exp(y)+1)),y=0..infinity);

$$\frac{457}{5042}\pi^{6} \tag{B.3}$$

# Appendix C

# Tables

Below, we get numerical values for the Neutrino emissivity which is plotted in Figure 7.8. Columns from left to right are as follows: total density, emissivity of the processes $\Lambda \to p e \bar{\nu}_e$, $\Sigma^- \to n e \bar{\nu}_e$, $\Sigma^- \to n \mu \bar{\nu}_\mu$, and the total emissivity. The density is given in fm$^{-3}$. The neutrino emissivity given in MeV s$^{-1}$ fm$^{-3}$, but we have extracted the common factors $(k_B T)^6 G_F^2$, so to get a final numerical value, these factors must be included.

```
0.2100D+00 0.000000D+00 0.000000D+00 0.000000D+00 0.000000D+00
0.2200D+00 0.000000D+00 0.000000D+00 0.000000D+00 0.000000D+00
0.2300D+00 0.000000D+00 0.000000D+00 0.000000D+00 0.000000D+00
0.2400D+00 0.000000D+00 0.000000D+00 0.000000D+00 0.000000D+00
0.2500D+00 0.000000D+00 0.000000D+00 0.000000D+00 0.000000D+00
0.2600D+00 0.000000D+00 0.000000D+00 0.000000D+00 0.000000D+00
0.2700D+00 0.000000D+00 0.000000D+00 0.000000D+00 0.000000D+00
0.2800D+00 0.000000D+00 0.156000D+07 0.156000D+07 0.312000D+07
0.2900D+00 0.000000D+00 0.154945D+07 0.154945D+07 0.309890D+07
0.3000D+00 0.000000D+00 0.153890D+07 0.153890D+07 0.307781D+07
0.3100D+00 0.000000D+00 0.152625D+07 0.152625D+07 0.305249D+07
0.3200D+00 0.000000D+00 0.151359D+07 0.151359D+07 0.302718D+07
0.3300D+00 0.000000D+00 0.149882D+07 0.149882D+07 0.299765D+07
0.3400D+00 0.000000D+00 0.148406D+07 0.148406D+07 0.296811D+07
0.3500D+00 0.000000D+00 0.146718D+07 0.146718D+07 0.293436D+07
0.3600D+00 0.000000D+00 0.145030D+07 0.145030D+07 0.290061D+07
0.3700D+00 0.000000D+00 0.143237D+07 0.143237D+07 0.286474D+07
0.3800D+00 0.000000D+00 0.141339D+07 0.141339D+07 0.282677D+07
0.3900D+00 0.000000D+00 0.139335D+07 0.139335D+07 0.278669D+07
0.4000D+00 0.000000D+00 0.137225D+07 0.137225D+07 0.274450D+07
0.4100D+00 0.000000D+00 0.135010D+07 0.135010D+07 0.270020D+07
```

```
0.4200D+00  0.000000D+00  0.132584D+07  0.132584D+07  0.265168D+07
0.4300D+00  0.000000D+00  0.130158D+07  0.130158D+07  0.260316D+07
0.4400D+00  0.000000D+00  0.127627D+07  0.127627D+07  0.255253D+07
0.4500D+00  0.000000D+00  0.125201D+07  0.125201D+07  0.250401D+07
0.4600D+00  0.000000D+00  0.122775D+07  0.122775D+07  0.245550D+07
0.4700D+00  0.000000D+00  0.120349D+07  0.120349D+07  0.240698D+07
0.4800D+00  0.000000D+00  0.117923D+07  0.117923D+07  0.235846D+07
0.4900D+00  0.000000D+00  0.115497D+07  0.115497D+07  0.230994D+07
0.5000D+00  0.000000D+00  0.113071D+07  0.000000D+00  0.113071D+07
0.5100D+00  0.000000D+00  0.110645D+07  0.000000D+00  0.110645D+07
0.5200D+00  0.000000D+00  0.108219D+07  0.000000D+00  0.108219D+07
0.5300D+00  0.000000D+00  0.105793D+07  0.000000D+00  0.105793D+07
0.5400D+00  0.000000D+00  0.103262D+07  0.000000D+00  0.103262D+07
0.5500D+00  0.000000D+00  0.100625D+07  0.000000D+00  0.100625D+07
0.5600D+00  0.000000D+00  0.979878D+06  0.000000D+00  0.979878D+06
0.5700D+00  0.000000D+00  0.952454D+06  0.000000D+00  0.952454D+06
0.5800D+00  0.000000D+00  0.923975D+06  0.000000D+00  0.923975D+06
0.5900D+00  0.000000D+00  0.895497D+06  0.000000D+00  0.895497D+06
0.6000D+00  0.000000D+00  0.865963D+06  0.000000D+00  0.865963D+06
0.6100D+00  0.000000D+00  0.836430D+06  0.000000D+00  0.836430D+06
0.6200D+00  0.237161D+07  0.807951D+06  0.000000D+00  0.317956D+07
0.6300D+00  0.229421D+07  0.781582D+06  0.000000D+00  0.307579D+07
0.6400D+00  0.221681D+07  0.755213D+06  0.000000D+00  0.297202D+07
0.6500D+00  0.213940D+07  0.728843D+06  0.000000D+00  0.286825D+07
0.6600D+00  0.206200D+07  0.702474D+06  0.000000D+00  0.276448D+07
0.6700D+00  0.198460D+07  0.676105D+06  0.000000D+00  0.266070D+07
0.6800D+00  0.190720D+07  0.649736D+06  0.000000D+00  0.255693D+07
0.6900D+00  0.183289D+07  0.624422D+06  0.000000D+00  0.245731D+07
0.7000D+00  0.175549D+07  0.598052D+06  0.000000D+00  0.235354D+07
0.7100D+00  0.168118D+07  0.572738D+06  0.000000D+00  0.225392D+07
0.7200D+00  0.160688D+07  0.547424D+06  0.000000D+00  0.215430D+07
0.7300D+00  0.153567D+07  0.523164D+06  0.000000D+00  0.205883D+07
0.7400D+00  0.146446D+07  0.498904D+06  0.000000D+00  0.196336D+07
0.7500D+00  0.139324D+07  0.474645D+06  0.000000D+00  0.186789D+07
0.7600D+00  0.132513D+07  0.451440D+06  0.000000D+00  0.177657D+07
0.7700D+00  0.125392D+07  0.427180D+06  0.000000D+00  0.168110D+07
0.7800D+00  0.118581D+07  0.403975D+06  0.000000D+00  0.158978D+07
0.7900D+00  0.111769D+07  0.380771D+06  0.000000D+00  0.149846D+07
0.8000D+00  0.104958D+07  0.357566D+06  0.000000D+00  0.140714D+07
0.8100D+00  0.981464D+06  0.334361D+06  0.000000D+00  0.131582D+07
0.8200D+00  0.907157D+06  0.309046D+06  0.000000D+00  0.121620D+07
```

0.8300D+00 0.839043D+06 0.285842D+06 0.000000D+00 0.112488D+07
0.8400D+00 0.770929D+06 0.262637D+06 0.000000D+00 0.103357D+07
0.8500D+00 0.705911D+06 0.240487D+06 0.000000D+00 0.946397D+06
0.8600D+00 0.000000D+00 0.000000D+00 0.000000D+00 0.000000D+00
0.8700D+00 0.000000D+00 0.000000D+00 0.000000D+00 0.000000D+00
0.8800D+00 0.000000D+00 0.000000D+00 0.000000D+00 0.000000D+00
0.8900D+00 0.000000D+00 0.000000D+00 0.000000D+00 0.000000D+00
0.9000D+00 0.000000D+00 0.000000D+00 0.000000D+00 0.000000D+00
0.9100D+00 0.000000D+00 0.000000D+00 0.000000D+00 0.000000D+00
0.9200D+00 0.000000D+00 0.000000D+00 0.000000D+00 0.000000D+00
0.9300D+00 0.000000D+00 0.000000D+00 0.000000D+00 0.000000D+00
0.9400D+00 0.000000D+00 0.000000D+00 0.000000D+00 0.000000D+00
0.9500D+00 0.000000D+00 0.000000D+00 0.000000D+00 0.000000D+00
0.9600D+00 0.000000D+00 0.000000D+00 0.000000D+00 0.000000D+00
0.9700D+00 0.000000D+00 0.000000D+00 0.000000D+00 0.000000D+00
0.9800D+00 0.000000D+00 0.000000D+00 0.000000D+00 0.000000D+00
0.9900D+00 0.000000D+00 0.000000D+00 0.000000D+00 0.000000D+00
0.1000D+01 0.000000D+00 0.000000D+00 0.000000D+00 0.000000D+00
0.1010D+01 0.000000D+00 0.000000D+00 0.000000D+00 0.000000D+00
0.1020D+01 0.000000D+00 0.000000D+00 0.000000D+00 0.000000D+00
0.1030D+01 0.000000D+00 0.000000D+00 0.000000D+00 0.000000D+00
0.1040D+01 0.000000D+00 0.000000D+00 0.000000D+00 0.000000D+00
0.1050D+01 0.000000D+00 0.000000D+00 0.000000D+00 0.000000D+00
0.1060D+01 0.000000D+00 0.000000D+00 0.000000D+00 0.000000D+00
0.1070D+01 0.000000D+00 0.000000D+00 0.000000D+00 0.000000D+00
0.1080D+01 0.000000D+00 0.000000D+00 0.000000D+00 0.000000D+00
0.1090D+01 0.000000D+00 0.000000D+00 0.000000D+00 0.000000D+00
0.1100D+01 0.000000D+00 0.000000D+00 0.000000D+00 0.000000D+00

# Bibliography

[1] **[Ph99]** A.C. Philips; The Physics of Stars, John Wiley & Sons (1999).

[2] **[KW90]** R.Kippenhahn A. Weigert; Stellar structure and Evolution, Springer-Verlag, (1990)380.

[3] **[Øs89]** E. Østgaard; Neutron Stars Rewiew, Fysisk Institutt, AVH, UNIT, 1989

[4] **[Pr94]** M. Prakash; Physics Reports 242 (1994) p 305.

[5] **[GS84]** Gaillard & Savage; Ann. Rev. Nucl. Part. Sci. (1984) 34: 359-364.

[6] **[Iw82]** N. Iwamoto; Ann.Phys. 144(1982).

[7] **[Gl00]** N.K. Glendenning; Compact Stars, Springer (2000).

[8] **[ST83]** S.L.Shapiro & S.A. Teukolsky; Black Holes, White Dwarfs and Neutron Stars, John Wiley & Sons (1983).

[9] **[LPP91]** J. M. Lattimer, C. J. Pethick, M. Prakash, P. Haensel; Phys. Rev. Lett. 66(1991)2703.

[10] **[PPL92]** M. Prakash, M. Prakash, J. M. Lattimer, C. J. Pethick; Astrophys. J. 390(1992)L77.

[11] **[BZ34]** W. Baade & F. Zwicky; Phys. Rev. 45(1934)138, Proc. Nat. Acad. Sci. 20(1934)259.

[12] **[OV39]** J.R. Oppenheimer & G.M. Volkoff; Phys. Rev. 55(1939)374.

[13] **[GGP62]** R. Giacconi, H. Gursky, F.R. Paolini & B.B. Rossi; Phys. Rev. Lett. 9(1962)439.

[14] **[HBP68]** A. Hewish, S.J. Bell, J.D.H. Pilkington, P.F Scott & R.A. Collins; Nature 217(1968)709.

[15] [**G68**] T. Gould; Nature 218(1968)731.

[16] [**HT75**] R.A. Hulse & J.H. Taylor; Astrophys. Journ. 195 (1975)l51.

[17] [**Pe92**] C.J. Pethick; Rev. Mod. Phys. 64(1992)1133.

[18] [**ØØ91**] T. Øvergård & E. Østgård; Astron. Astrophys. 243(1991)412.

[19] [**Ca74**] V. Canuto; Ann. Rev. Astron. Astrophys. 12(1974)167.

[20] [**Rø01**] T. Røste; Nøytronstjerners maksimale masse og radius, Hovedoppgave, Institutt for fysikk, NTNU (2001).

[21] [**Mø83**] T. Mølnvik; Nøytronstjerners maksimale masse og treghetsmoment, Hovedoppgave, NLTH (1983).

[22] [**ABS75**] R. Adler, M. Bazin & M. Schiffer; Introduction to General Relativity, McGraw-Hill Book Company, 2nd edition (1975).

[23] [**Dy91**] B. Dybvik; Tilstandsligninger, masse og radius for nøytronstjerner, Hovedoppgave, UNIT/NTH (1991).

[24] [**BG99**] L. Bergström & A. Goobar; Cosmology and Particle Astrophysics, John Wiley & Sons (1999).

[25] [**Be89**] M. V. Berry; Principles of cosmology and gravitation, Institute of Physics Publishing Bristol, Philadelphia (1989).

[26] [**HH00**] H. Heiselberg, M. Hjorth-Jensen; Phys. Rep. 328(2000)237.

[27] [**KN86**] D.B. Kaplan, A.E. Nelson; Phys. Lett. B 175(1986)57; B 179(1986)409(E).

[28] [**PBP97**] M. Prakash, I. Bombaci, M. Prakash, P. J. Ellis, J.M. Lattimer, R. Knorren; Phys Rep. 280(1997)1-77.

[29] [**MBB85**] C. Mahaux, P.F. Bortingnon, R.A. BrogliaC.H. Dasso, Phys. Rep. 120 (1985) 1.

[30] [**MST90**] A.B. Migdal, E.E. Saperstein, M.A. Troitsky, D.N. Voskresensky, Phys. Rep. 192 (1990) 179.

[31] [**MSS96**] R. Machleidt, F. Sammarruca, Y. Song, Phys. Rev. C 53 (1996) R1483.

[32] [**WSS95**] R.B. Wiringa, V.G.J. Stoks, R. Schiavilla, Phys. Rev. C 51 (1995) 38.

[33] **[SKT94]** V.G.J. Stoks, R.A.M. Klomp, C.P.F. Terheggen, J.J. de Swart, Phys. Rev. C 49 (1994) 2950.

[34] **[NHK97]** A. Nogga, D. Hüber, H. Kamada, W. Glöcke, Phys Lett. B 409 (1997) 19.

[35] **[SBG98]** H.Q. Song, M. Baldo, G. Giansiracusa, U. Lombardo, Phys. Rev. Lett. 81 (1998) 1584.

[36] **[SBG97]** H.Q. Song, M. Baldo, G. Giansiracusa, U. Lombardo, Phys. Rev. Lett. B 411 (1997) 237.

[37] **[APR98]** A. Akmal, V.R. Pandharipande, D.G. Ravenhall, Phys. Rev. C 58 (1998) 1804.

[38] **[Da81]** B.D. Day, Phys. Rev. C 24 (1981) 1203.

[39] **[Fr75]** J.L. Friar, Phys. Rev. C 12 (1975) 695.

[40] **[AP97]** A. Akmal, V.R. Pandharipande, Phys. Rev C 56 (1997) 2261.

[41] **[BK96]** V. Kalogera, G. Baym, Astrophys. J 470 (1996) L61.

[42] **[RSY98]** Th.A. Rijken, V.G.J. Stoks, Y. Yamamoto, Phys. Rev. C 59 (1998) 21.

[43] **[SR99]** V.G.J. Stoks, T.A. Rijken, Phys. Rev. C 59 (1999) 3009.

[44] **[SL99]** V.G.J. Stoks, T.S.H. Lee, Phys. Rev. C 60 (1999) 024006.

[45] **[VPR00]** I. Vidaña, A. Polls, A. Ramos, L. Engvik, M. Hjorth-Jensen, Phys. Rev. C 62 (2000) 035801.

[46] **[FM79]** B.L. Friman, O.V. Maxwell, Astrophys. J. 232 (1979) 451.

[47] **[EEO96]** Ø. Elgarøy, L. Engvik, E. Osnes, F.V. De Blasio, M. Hjorth-Jensen, G. Lazzari, Phys. Rev. Lett. 76 (1996) 12.

[48] **[YKG01]** D.G. Yakovlev, A.D. Kaminker, O.Y. Gnedin, P. Haensel, Phys. Rep. 354 (2001) 1-155.

Printed in Great Britain
by Amazon